EXAM

ON THE

INTEGRAL CALCULUS.

BY JAMES HANN,

MATHEMATICAL MASTER OF KING'S COLLEGE SCHOOL, LONDON

HONORARY MEMBER OF THE PHILOSOPHICAL SOCIETY OF NEWCASTLE-UPON-TYNE,
JOINT AUTHOR OF MECHANICS FOR PRACTICAL MEN,
AUTHOR OF THEORY OF BRIDGES, JOINT AUTHOR OF NAUTICAL TABLES,
AUTHOR OF A TREATISE ON THE STEAM ENGINE,
AUTHOR OF THEORETICAL AND PRACTICAL MECHANICS,
AND AUTHOR OF PLANE AND SPHERICAL
TRIGONOMETRY.

LONDON:
JOHN WEALE, 59, HIGH HOLBORN.

1850.

PREFACE.

As this work contains a great number of Integrals fully worked out, the Author hopes that it will considerably facilitate the progress of those who are entering on this branch of study, by showing them almost all the artifices that are used in those branches that come within its scope.

The works that have been consulted are those of Peacock, Gregory, Hall, De Morgan, Young, and various mathematical periodicals, also the excellent little work on the Calculus by Mr Tate, which, like all the productions of that eminent writer, abounds with useful information, apart from the able manner in which he has treated the first principles.

Where integration by parts is used, the whole process is put down, but the student should endeavour as soon as possible to acquire the facility of running off the quantities without writing down all the intermediate steps

$$\int (du \pm dv \pm dw) = u \pm v \pm w$$

$$\int \frac{dx}{x \log a} = \log x, \quad \int \frac{dx}{x} = \log x$$

$$a \, a^u \, dx \qquad a^x$$

$$\cos u \, du \quad \sin$$

$$\int -\sin u \, du \quad \cos u$$

$$\int -\csc^2 a \cos u \, du \quad \csc u$$

$$\sec^2 u \, du \quad \tan u$$

$$\int -\csc^2 u \, du \quad \cot u$$

$$\int \frac{du}{1 - u^2} \quad \arcsin u$$

15. $\int \frac{du}{u \, u^2} \quad \arccos u$ [?]

$\frac{}{u^2} \quad \arccos u$

17. $\int \frac{du}{1 + u} \quad \arctan u$

18. $\int -\frac{du}{\sqrt{1 + u^2}} \quad \arccot u$

$\int \frac{du}{x\sqrt{x^2 - 1}} \quad \arcsec u$

20.

EXAMPLES

ON THE

INTEGRAL CALCULUS.

CHAPTER I.

ELEMENTARY INTEGRALS TO BE COMMITTED TO MEMORY.

(1.) $\int \frac{d\,x}{x} = \log x.$

(2.) $\int \frac{d\,x}{a+b\,x^2} = \frac{1}{\sqrt{ab}} \tan^{-1}\left(x \sqrt{\frac{b}{a}} \right),$ or $\int \frac{d\,x}{a^2+x^2}$

$= \frac{1}{a} \tan^{-1} \frac{x}{a}.$

(3.) $\int \frac{d\,x}{\sqrt{a^2-x^2}} = \sin^{-1} \frac{x}{a}.$

(4.) $\int \frac{-d\,x}{\sqrt{a^2-x^2}} = \cos^{-1} \frac{x}{a}.$

(5.) $\int \frac{d\,x}{\sqrt{x^2 \pm a^2}} = \log \frac{(x+\sqrt{x^2 \pm a^2})}{a}.$

(6.) $\int \frac{d\,x}{\sqrt{2\,a\,x-x^2}} = \text{vers}^{-1} \frac{x}{a}.$

(7.) $\int \frac{d\,x}{\sqrt{x^2 \pm 2\,a\,x}} = \log\,(x \pm a + \sqrt{x^2 \pm 2\,a\,x}.$

(8.) $\int \frac{d\,x}{x \sqrt{x^2-a^2}} = \frac{1}{a} \sec^{-1} \frac{x}{a}.$

(9.) $\int \dfrac{d x}{x\sqrt{a^2 \pm x^2}} = \dfrac{1}{a} \log \dfrac{x}{a + \sqrt{a^2 \pm x^2}}$

(10.) $\int \dfrac{a\, d x}{x^2 - a^2} = \dfrac{1}{2} \log \left(\dfrac{x-a}{x+a} \right).$

(11.) $\int a^x d x = \dfrac{a^x}{\log a}.$

(12.) $\int e^{ax}\, d x = \dfrac{e^{ax}}{a}.$

(13.) $\int \dfrac{d\theta}{\sin \theta} = \log \tan \dfrac{\theta}{2}$

(14.) $\int \dfrac{d\theta}{\cos \theta} = \log \tan \left\{ \dfrac{\pi}{4} + \dfrac{\theta}{2} \right\}$

(15.) $\int \dfrac{d\theta}{\sin \theta \cos \theta} = \log \tan \theta.$

Examples.

(1.) $\int \dfrac{d x}{1 + 5 x^2} = \dfrac{1}{\sqrt{5}} \tan^{-1} (x\sqrt{5}).$

(2.) $\int \dfrac{d x}{x^2 (a + b x)}.$ Let $x = \dfrac{1}{z}$, then the integral is re-

duced to $\int -\dfrac{z\, dz}{az + b} = -\dfrac{z}{a} + \dfrac{b}{a^2} \log (az + b)$

$\qquad\qquad = -\dfrac{1}{a x} + \dfrac{b}{a^2} \log \left(\dfrac{a + b x}{x} \right).$

(3.) $\int \left\{ \dfrac{a\, d x}{x} + \dfrac{b\, d x}{x^2} + \dfrac{c\, d x}{x^3} + \dfrac{e\, d x}{x^4} \right\} = a \log x - \dfrac{b}{x}$

$\qquad\qquad - \dfrac{c}{2 x^2} - \dfrac{e}{3 x^3}$

(4.) $\int (1 + x^2)(1 + x)\, x\, d x = \int (1 + x + x^2 + x^3)\, x\, d x$

$$= \int (x + x^2 + x^3 + x^4)\, dx = \frac{x^2}{2} + \frac{x^3}{3} + \frac{x^4}{4} + \frac{x^5}{5}.$$

(5.) $$\int \frac{(1+x)^2(1-x)\,dx}{x^2} = \int \frac{1+x-x^2-x^3\,dx}{x^3} = \int \left(\frac{1}{x^3} + \frac{1}{x} \right.$$

$$\left. -1-x \right) dx = \log x - x - \frac{x^2}{2} - \frac{1}{x} = \frac{x \log x^3 - 2\,x^2 - x^3 - 2}{2\,x}$$

(6.) $$\int \frac{x^4\,dx}{x^2+1} = \int \left\{ x^2 - 1 + \frac{1}{x^2+1} \right\} dx = \frac{x^3}{3} - x + \tan^{-1} x.$$

(7.) $$\int du = \int \frac{x^3\,dx}{(x+2)^2}. \quad \text{Let } x+2 = z;\ dx = dz$$

$$\therefore\ du = \frac{\{z-2\}^3 dz}{(z)^2} = \frac{(z^3 - 6\,z^2 + 12\,z - 8)\,dz}{z^2}$$

$$= \left(z - 6 + \frac{12}{z} - \frac{8}{z^2} \right) dz,\ \therefore\ u = \frac{z^2}{2} - 6\,z + 12 . \log z + \frac{8}{z},$$

$$= \frac{z^3 - 12\,z^2 + z . \log (z^{24}) + 16}{2\,z}$$

$$= \frac{(x+2)^3 - 12\,(x+2)^2 + (x+2)\log(x-2)^{24} + 16}{2\,(x+2)}.$$

(8.) $$\int \frac{x^3\,dx}{x^2+1} = \int \left(x^3 - x + \frac{x}{x^2+1} \right) dx$$

$$= \frac{x^4}{4} - \frac{x^2}{2} + \log(\sqrt{x^2+1}).$$

(9.) $$\int \frac{5\,dx}{2\,x^4 + 3\,x^2}. \quad \text{Let } x = \frac{1}{z}, \text{ then the integral is}$$

reduced to $$\int \frac{-5\,z^2\,dz}{2+3\,z^2}$$

$$= -\frac{1}{3} \int \frac{15\,z^2\,dz}{2+3\,z^2} = -\frac{1}{3} \int \left(5\,dz - \frac{10\,dz}{3\,z^2+2} \right)$$

$$= \int \left(\frac{10}{3} \cdot \frac{dz}{3\,z^2+2} - \frac{5\,dz}{3} \right)$$

$$= \frac{1}{3} \left\{ \left(\frac{5\sqrt{2}}{\sqrt{3}} \right) \tan^{-1} \left(\sqrt{\frac{3}{2}} \cdot z \right) - 5\,z \right\}$$

$$= \frac{1}{3} \left(5\sqrt{\frac{2}{3}} \tan^{-1} \left(\sqrt{\frac{3}{2\,x^2}} \right) - \frac{5}{x} \right).$$

(10.) $\displaystyle\int \frac{x\sqrt{x}\,dx}{1+\sqrt{x}} = \int \left(x - \sqrt{x} + 1 - \frac{1}{1+\sqrt{x}} \right) dx$

$$= \frac{x^2}{2} - \frac{2\,x^{\frac{3}{2}}}{3} + x - \int \frac{dx}{1+\sqrt{x}}.$$

And $\displaystyle\int \frac{dx}{1+\sqrt{x}} =$ (by substituting $\sqrt{x}=z$)

$$2 \int \frac{z\,dz}{1+z} = 2 \int \left(dz - \frac{dz}{1+z} \right),$$

$$= 2z - \log(1+z)^2 = 2\sqrt{x} - \log(1+\sqrt{x})^2;$$

$\therefore$ the integral is $= \dfrac{x^2}{2} - \dfrac{2\,x^{\frac{3}{2}}}{3} + x - 2\,(x)^{\frac{1}{2}} + \log(1+\sqrt{\;})^2.$

(11.) $\displaystyle\int \frac{x^2\sqrt{x}\,.\,dx}{1+x}.$ Let $\sqrt{x}=z$, then $x=z^2$

$$dx = 2\,z\,dz;$$

$$\therefore \int \frac{x^2\sqrt{x}\,.\,dx}{1+x} = \int \frac{z^5}{1+z^2}\,.\,2\,z\,dz = 2\int \frac{z^6\,dz}{1+z^2}$$

$$= 2\int \left\{ z^4\,dz - z^2\,dz + dz - \frac{dz}{1+z^2} \right\}$$

$$+ \frac{2\,z^5}{5} - \frac{2\,z^3}{3} + 2\,z - 2\,\tan^{-1}z$$

$$= \frac{2}{5}\,x^{\frac{5}{2}} - \frac{2}{3}\,x^{\frac{3}{2}} + 2\,x^{\frac{1}{2}} - 2\,\tan^{-1}x^{\frac{1}{2}}$$

$$(12.) \int \frac{dx}{1+x+x^2} = \int \frac{dx}{(x+\frac{1}{2})^2+\frac{3}{4}}$$

$$= \frac{2}{\sqrt{3}} \tan^{-1}\left\{\frac{2x+1}{\sqrt{3}}\right\} \text{ (an useful integral)}$$

$$\int \frac{dx}{a+bx+cx^2} = \frac{1}{c}\int \frac{dx}{x^2+\frac{b}{c}x+\frac{a}{c}} = 2\int \frac{d(2cx+b)}{(2cx+b)^2+(4ac-b^2)}$$

$$= \frac{2}{\sqrt{4ac-b^2}} \tan^{-1}\left(\frac{2cx+b}{\sqrt{4ac-b^2}}\right).$$

Examples for Practice.

$$(1.) \int (a+bx)dx = \frac{(a+bx)^2}{2b}.$$

$$(2.) \int x(a+bx^2)^3 dx = \frac{(a+bx^2)^4}{8b}.$$

$$(3.) \int \frac{(a+bx^2+cx^3)(3cx^2+2bx)dx}{4} = \frac{(a+bx^2+cx^3)^2}{8}$$

$$(4.) \int \frac{\{(x+\sqrt{x})(2\sqrt{x}+1)\}dx}{\sqrt{x}} = (x+\sqrt{x})^2.$$

$$(5.) \int_b^0 (a+bx)xdx = \frac{3ab^2+2b^4}{6}.$$

$$(6) \int \frac{(\sqrt{x^2+4}+x)dx}{\sqrt{x^2+4}} = (x+\sqrt{x^2+4}).$$

$$(7.) \int \frac{(\sqrt{x^2+4}+x)^2dx}{\sqrt{x^2+4}} = \frac{1}{2}(x+\sqrt{x^2+4})^2.$$

$$(8.) \int (a+bx^2)^2.2xb\,dx = \frac{(a+bx^2)^3}{3}.$$

(9.) $\displaystyle\int \frac{3\,dx}{\sqrt{x^2+3}\cdot(\sqrt{x^2+3}-x)} = x + \sqrt{x^2+3}.$

(10.) $\displaystyle\int \frac{(2x+1)\,dx}{\sqrt{x^2+x}} = \frac{\sqrt{x^2+x}}{\frac{1}{2}} = 2\sqrt{x^2+x}\cdot$

(11.) $\displaystyle\int \frac{(3x+2)\,dx}{\sqrt{x+1}} = 2\sqrt{x^3+x^2}.$

(12.) $\displaystyle\int \frac{2x\,dx}{(x^2+1)^2} = -\frac{1}{x^2+1}\cdot$

(13.) $\displaystyle\int \frac{dx}{a+bx} = \frac{1}{b}\log\sqrt{a+bx}.$

(14.) $\displaystyle\int \frac{(2x+1)\,dx}{\sqrt{x^2+x+1}} = 2\sqrt{x^2+x+1}.$

(15.) $\displaystyle\int \frac{\left(x^3 + \dfrac{3\,c}{4\,d}x^2 + \dfrac{b}{2\,d}x\right)dx}{\sqrt{a+bx^2+cx^3+dx^4}} = \frac{\sqrt{a+bx^2+cx^3+dx^4}}{2d}$

(16.) $\displaystyle\int \frac{(9x^8 + 8x^7)\,dx}{(x^9+x^8)^{\frac{3}{5}}} = \frac{5\,(x^9+x^8)^{\frac{2}{5}}}{2}.$

(17.) $\displaystyle\int \frac{\left(x^{n-1} + \dfrac{n-1}{n}x^{n-2}\right)dx}{(x^n+x^{n-1})^{\frac{p}{q}}} = \frac{q.(x^n+x^{n-1})^{\frac{q-p}{q}}}{n.(q-p)}\cdot$

(18.) $\displaystyle\int \frac{\left(x + \dfrac{b}{2c}\right)dx}{\sqrt{a+bx+cx^2}} = \frac{\sqrt{a+bx+cx^2}}{c}.$

(19.) $\displaystyle\int \frac{\left(x + \dfrac{b}{2c}\right)dx}{(a+bx+cx^2)^{\frac{m}{n}}} = \frac{n.(a+bx+cx^2)^{\frac{n-m}{n}}}{(n-m).2c}.$

CHAPTER II.

RATIONAL FRACTIONS.

(1.) To integrate $du = \dfrac{x\,dx}{(x+2)(x+3)^2}$.

Let $\dfrac{x}{(x+2)(x+3)^2} = \dfrac{A}{(x+3)^2} + \dfrac{B}{(x+3)} + \dfrac{P}{(x+2)}$,

$\therefore\ x = A(x+2) + B(x+3)(x+2) + P(x+3)^2$.

Let $x = -3$,

$\qquad \therefore\ -3 = A(2-3) = -A, \quad \therefore\ A = 3$,

$x - 3(x+2) = B(x+3)(x+2) + P(x+3)^2$

$-2(x+3) = B(x+3)(x+2) + P(x+3)^2$

$\qquad \therefore\ -2 = B(x+2) + P(x+3)$.

Let $x = -3$,

$\qquad -2 = B(2-3) = -B, \quad \therefore\ B = 2$.

Let $x = -2$,

$\qquad -2 = P(3-2) = P, \quad \therefore\ P = -2$,

$\therefore\ \dfrac{x}{(x+2)(x+3)^2} = \dfrac{3}{(x+3)^2} + \dfrac{2}{(x+3)} - \dfrac{2}{(x+2)}$,

$\therefore\ \displaystyle\int \dfrac{x\,dx}{(x+2)(x+3)^2}$

$\qquad = \displaystyle\int \dfrac{3\,dx}{(x+3)^2} + 2\int \dfrac{dx}{(x+3)} - 2\int \dfrac{dx}{(x+2)}$

$\qquad = -\dfrac{3}{(x+3)} + \log\left(\dfrac{x+3}{x+2}\right)^2$.

(2.) To find $\int \dfrac{x^2 dx}{(x+2)^2 (x+4)^2}.$ Let

$$\dfrac{x^2}{(x+4)^2(x+2)^2} = \dfrac{A}{(x+4)^2} + \dfrac{B}{(x+4)} + \dfrac{C}{(x+2)^2} + \dfrac{D}{x+2}$$

$$x^2 = A(x+2)^2 + B(x+4)(x+2)^2 + C(x+4)^2 + D(x+2)(x+4)^2.$$

Let $x+2=0$, $\therefore x=-2$, $x^2=4$, and $C(x+4)^2=4C$,

$$\therefore 4=4C, \quad \therefore C=1.$$

Let $x+4=0$, $\therefore x=-4$, $x^2=16$, and $A(x+2)^2=4A$,

$$\therefore A=4,$$

$$\therefore x^2 - (x+4)^2 - 4(x+2)^2 = -4\{x^2+6x+8\}$$

$$= -4\{(x+2)(x+4)\},$$

$$= B(x+4)(x+2)^2 + D(x+2)(x+4)^2,$$

$$\therefore -4 = B(x+2) + D(x+4).$$

Let $x=-2$, $\quad \therefore -4 = 2D$, $\quad \therefore D=-2.$

Let $x=-4$, $\quad \therefore -4 = -2B$, $\quad \therefore B=2.$

The fraction reduced becomes, therefore,

$$\dfrac{4}{(x+4)^2} + \dfrac{2}{(x+4)} + \dfrac{1}{(x+2)^2} - \dfrac{2}{(x+2)};$$

and its integral is, therefore,

$$\int \dfrac{4\,dx}{(x+4)^2} + \int \dfrac{dx}{(x+2)^2} + \log\left(\dfrac{x+4}{x+2}\right)^2.$$

$$\int \dfrac{4\,dx}{(x+4)^2} = \int 4\,dx\,(x+4)^{-2};$$

$$= \frac{-4}{(x+4)}.$$

In the same way,

$$\int \frac{dx}{(x+2)^2} = \frac{-1}{x+2},$$

and

$$-\left\{ \frac{4}{x+4} + \frac{1}{x+2} \right\} = -\frac{5x+12}{x^2+6x+8};$$

therefore the complete integral is

$$-\frac{5x+12}{x^2+6x+8} + \log \left(\frac{x+4}{x+2} \right)^2$$

(3.) $\displaystyle\int \frac{2\,x\,dx}{(x^2+1)(x^2+3)}.$

Let $\displaystyle \frac{2x}{(x^2+1)(x^2+3)} = \frac{Ax}{x^2+1} + \frac{Bx}{x^2+3}$;

$$\therefore\ 2 = A(x^2+3) + B(x^2+1),\ \ \therefore\ A+B=0,$$

$$3A+B=2,\ \ \therefore\ A=1,\ B=-1,$$

$$\int \frac{2\,x\,dx}{(x^2+1)(x^2+3)} = \int \left(\frac{x\,dx}{x^2+1} - \frac{x\,dx}{x^2+3} \right)$$

$$= \log \sqrt{\frac{x^2+1}{x^2+3}}$$

(4.) $\displaystyle du = \frac{x^2 dx}{(x^2+1)(x^2+4)}.$

Let $\dfrac{x^2}{(x^2 + 1)(x^2 + 4)} = \dfrac{A}{(x^2 + 1)} + \dfrac{B}{(x^2 + 4)}$

$$x^2 = A(x^2 + 4) + B(x^2 + 1).$$

Let $x = \sqrt{-1}$, or $x^2 = -1$,

$$\therefore -1 = 3A, \quad \therefore A = -\frac{1}{3},$$

$$\therefore x^2 + \frac{1}{3}(x^2 + 4) = B(x^2 + 1),$$

$$\therefore \frac{4(x^2 + 1)}{3} = B(x^2 + 1), \quad \therefore B = \frac{4}{3}$$

$$\int \frac{x^2\,dx}{(x^2 + 1)(x^2 + 4)} = -\frac{1}{3}\int \left\{\frac{dx}{(x^2 + 1)} - \int\frac{4\,dx}{(x^2 + 4)}\right\}$$

$$= \frac{1}{3}\int \left\{\frac{dx}{(1 + x^2)} - \frac{dx}{\left(1 + \dfrac{x^2}{4}\right)}\right\}$$

$$= \frac{1}{3}\left\{2\tan^{-1}\frac{x}{2} - \tan^{-1}x\right\}.$$

Or thus, $\displaystyle \int \frac{x^2\,dx}{(x^2 + 1)(x^2 + 4)} = \frac{1}{3}\int \frac{3x^2\,dx}{(x^2 + 1)(x^2 + 4)}$

$$= \frac{1}{3}\int \frac{\{4x^2 + 4 - (x^2 + 4)\}\,dx}{(x^2 + 1)(x^2 + 4)}$$

$$= \frac{1}{3}\int \left(\frac{4\,dx}{(x^2 + 4)} - \frac{dx}{x^2 + 1}\right)$$

$$= \frac{1}{3}\left(2\tan^{-1}\frac{x}{2} - \tan^{-1}x\right)$$

(5.) $du = \dfrac{(3x^2 + x - 2)\,dx}{(x - 1)^3(x^2 + 1)}.$

Let $\dfrac{(3x^2 + x - 2)}{(x-1)^3(x^2+1)}$

$$= \frac{A}{(x-1)^3} + \frac{B}{(x-1)^2} + \frac{C}{(x-1)} + \frac{P}{(x^2+1)}$$

$$3x^2 + x - 2 = A(x^2+1) + B(x-1)(x^2+1) + C(x-1)^2(x^2+1)$$
$$+ P(x-1)^3.$$

Let $x = 1$ $\therefore 2 = 2A,$ $\therefore A = 1,$

$$3x^2 + x - 2 - (x^2+1) = B(x-1)(x^2+1) + C(x-1)^2(x^2+1)$$
$$+ P(x-1)^3,$$

$$\therefore (2x+3)(x-1) = B(x-1)(x^2+1) + C(x-1)^2(x^2+1)$$
$$+ P(x-1)^3,$$

$$\therefore (2x+3) = B(x^2+1) + C(x-1)(x^2+1) + P(x-1)^2.$$

Let $x = 1,$

$$\therefore 5 = 2B, \qquad \therefore B = \frac{5}{2},$$

$$2x + 3 - \frac{5}{2}(x^2+1)$$

$$= -\frac{5x^2 - 4x - 1}{2}$$

$$= -\frac{(5x+1)(x-1)}{2} = C(x^2+1)(x-1) + P(x-1)^3$$

$$-\frac{5x+1}{2} = C(x^2+1) + P(x-1); \text{ if } x = 1,$$

$$\therefore -3 = 2C, \qquad \therefore C = -\frac{3}{2}$$

$$-\frac{5x+1}{2} + \frac{3}{2}(x^2+1) = P(x-1)$$

$$\frac{3x^2 - 5x + 2}{2} = P(x-1)$$

$$\frac{(3x-2)(x-1)}{2} = P(x-1),$$

$$\therefore P = \frac{3x-2}{2} = \left(\frac{3x}{2} - 1\right)$$

$$\therefore \int \frac{(3x^2 + x - 2)dx}{(x-1)^3(x^2+1)} = \int \frac{dx}{(x-1)^3} + \frac{5}{2}\int \frac{dx}{(x-1)^2}$$

$$- \frac{3}{2}\int \frac{dx}{(x-1)} + \frac{3}{2}\int \frac{xdx}{(x^2+1)}$$

$$- \int \frac{dx}{(x^2+1)}$$

$$= - \frac{1}{2(x-1)^2} - \frac{5}{2}\frac{1}{(x-1)} - \frac{3}{2}\log(x-1)$$

$$+ \frac{3}{2}\cdot\frac{1}{2}\int \frac{2x}{(x^2+1)} - \tan^{-1}x$$

$$= - \frac{1}{2(x-1)^2} - \frac{5}{2}\cdot\frac{1}{x-1} + \frac{3}{2}\log\frac{\sqrt{x^2+1}}{(x-1)} - \tan^{-1}x$$

(6.) $du = \dfrac{(1-x+x^2)\,dx}{1+x+x^2+x^3} = \dfrac{(1-x+x^2)\,dx}{(1+x)(1+x^2)}.$

Let $\dfrac{1-x+x^2}{(1+x)(1+x^2)} = \dfrac{A}{(1+x)} + \dfrac{B}{(1+x^2)},$

$$\therefore 1 - x + x^2 = A(1+x^2) + B(1+x)$$

$$x = -1,$$

then $3 = 2A,$ $\therefore A = \dfrac{3}{2},$

$$1 - x + x^2 - \frac{3}{2}(1+x^2) = B(1+x)$$

$$= - \frac{x^2 + 2x + 1}{2} = - \frac{(1+x)^2}{2} = B(1+x),$$

$$\therefore \ B = -\ \frac{1+x}{2}$$

$$= -\frac{1}{2} - \frac{x}{2},$$

$$\therefore \int \frac{(1-x+x^2)\,dx}{1+x+x^2+x^3}$$

$$= \int \left(\frac{3}{2} \cdot \frac{1}{(1+x)} - \frac{1}{2} \cdot \frac{1}{1+x^2} - \frac{1}{2} \cdot \frac{x}{1+x^2} \right) dx$$

$$= \frac{1}{2} \left(3 \log (1+x) - \tan^{-1} x - \frac{1}{2} \log (1+x^2)^{\frac{1}{2}} \right)$$

$$= \frac{1}{2} \left(\log \frac{(1+x)^3}{\sqrt[4]{1+x^3}} - \tan^{-1} x \right)$$

(7.) $du = \dfrac{dx}{x\,(1+x)^2\,(1+x+x^2)}.$

Let $\dfrac{1}{x\,(x+1)^2\,(1+x+x^2)}$

$$= \frac{A}{x} + \frac{B}{(1+x)^2} + \frac{C}{(1+x)} + \frac{P}{1+x+x^2},$$

$\therefore \ 1 = A\,(1+x)^2\,(1+x+x^2) + B\,(1+x+x^2)\,x$

$\qquad + C\,(1+x)\,(1+x+x^2)\,x + P\,(1+x)^2\,x.$

Let $x = 0,$ $\quad \therefore\ A = 1,$

$$1 - (1+x)^2\,(1+x+x^2)$$

$$= B\,(1+x+x^2)\,x + C\,(1+x)\,(1+x+x^2)x + P\,(1+x)^2 x$$

$$\therefore \ -(3 + 4x + 3x^2 + x^4) = B\,(1+x+x^2)$$

$$+ C\,(1+x)\,(1+x+x^2) + P\,(1+x)^2.$$

Let $x = -1$, $\therefore\ B = -1$,

$$\therefore\ -(3 + 4x + 3x^2 + x^3) + (1 + x + x^2) =$$

$$C(1 + x)(1 + x + x^2) + P(1 + x)^2 \text{ or}$$

$$-(2 + 3x + 2x^2 + x^3) = C(1+x)(1+x+x^2) + P(1+x)^2$$

$$-(2 + x + x^2)(1 + x) = C(1+x)(1+x+x^2) + P(1+x)^2$$

$$-(2 + x + x^2) = C(1 + x + x^2) + P(1 + x)$$

$$x = -1, \qquad \therefore\ C = -2,$$

$$-(2 + x + x^2) + 2(1 + x + x^2) = P(1 + x)$$

$$x(1 + x) = P(1 + x), \qquad \therefore\ P = x,$$

$$\therefore \int \frac{dx}{x(1 + x)^2(1 + x + x^2)} =$$

$$\int \left(\frac{dx}{x} - \frac{dx}{(1 + x)^2} - \frac{2\,dx}{(1 + x)} + \frac{x\,dx}{1 + x + x^2} \right)$$

$$= \frac{1}{(1 + x)} - 2\log(1 + x) + \int \frac{(1 + x + 2x^2)\,dx}{x + x^2 + x^3}$$

$$= \frac{1}{1 + x} - 2\log(1 + x) +$$

$$\frac{1}{2} \int dx \left(\frac{2x + 3x^2 + 4x^3}{x^2 + x^3 + x^4} - \frac{x^2}{x^2 + x^3 + x^4} \right)$$

$$= \frac{1}{1 + x} + \log \frac{\sqrt{x^2 + x^3 + x^4}}{(1 + x)^2} - \int \frac{x^2\,dx}{x^2 + x^3 + x^4} \cdot \frac{1}{2}$$

$$\int\frac{x^2\,dx}{2\,(x^2 + x^3 + x^4)} = \int\frac{2\,dx}{4 + 4x + 4x^2} = \int\frac{2\,dx}{3 + (2x + 1)^2}$$

$$= \frac{2}{3}\int\frac{dx}{1 + \dfrac{(2x + 1)^2}{3}}; \text{ let } \frac{2x + 1}{\sqrt{3}} = z,$$

$$\therefore\ dx = \frac{\sqrt{3}\,.\,dz}{2},$$

$$\therefore\ \frac{2}{3}\int\frac{dx}{1 + \dfrac{(2x - 1)^2}{3}} = \frac{2}{3}\cdot\frac{\sqrt{3}}{2}\cdot\int\frac{dz}{1 + z^2} = \frac{1}{\sqrt{3}}\cdot\int\frac{dz}{1 + z^2}$$

$$= \frac{1}{\sqrt{3}}\tan^{-1} z = \frac{1}{\sqrt{3}}\tan^{-1}\frac{2x + 1}{\sqrt{3}}.$$

$$\therefore\int\frac{dx}{x\,(1 + x)^2\,(1 + x + x^2)} = \frac{1}{1 + x} + \log\frac{\sqrt{x^2 + x^3 + x^4}}{(1 + x)^3}$$

$$-\ \frac{1}{\sqrt{3}}\tan^{-1}\frac{2x + 1}{\sqrt{3}}.$$

Or thus,

$$\frac{1}{x\,(1 + x)^2\,(1 + x + x^2)}$$

$$= \frac{1 + x + x^2 - x - x^2}{x(1 + x)^2\,(1 + x + x^2)} = \frac{1}{x(1 + x)^2} - \frac{1}{(1 + x)\,(1 + x + x^2)}$$

$$= \frac{1 + x - x}{x\,(1 + x)^2} - \left\{\frac{1 + x + x^2 - x - x^2}{(1 + x)\,(1 + x + x^2)}\right\}$$

$$= \frac{1}{x(1+x)} - \frac{1}{(1+x)^2} - \frac{1}{1+x} + \frac{x}{(1+x+x^2)}$$

$$= \frac{1+x-x}{x(1+x)} - \frac{1}{1+x)^2} - \frac{1}{(1+x)} + \frac{x}{1+x+x^2}$$

$$= \frac{1}{x} - \frac{2}{1+x} - \frac{1}{(1+x)^2} + \frac{x}{(1+x+x^2)},$$

$$\therefore \int \frac{dx}{x(1+x)^2(1+x+x^2)} =$$

$$\frac{1}{x+1} + \log\left\{\frac{x}{(x+1)^2}\right\} + \int \frac{x\,dx}{(1+x+x^2)}.$$

$$\text{And } \int \frac{x\,dx}{(1+x+x^2)} = \int \left\{\frac{(x+\frac{1}{2}) - \frac{1}{2}}{(x+\frac{1}{2})^2 + \frac{3}{4}}\right\} dx$$

$$= \log\sqrt{x^2+x+1} - \frac{1}{\sqrt{3}} \tan^{-1}\left(\frac{2x+1}{\sqrt{3}}\right),$$

$$\therefore \int \frac{dx}{x(1+x)^2(1+x+x^2)} =$$

$$\frac{1}{x+1} + \log\frac{\sqrt{x^2+x^1+x^1}}{(x+1)^2} - \frac{1}{\sqrt{3}} \tan^{-1}\left(\frac{2x+1}{\sqrt{3}}\right).$$

$$(8.)\ \int \frac{dx}{x^8 + x^7 - x^4 - x^3} = \int \frac{dx}{x^3(x+1)^2(x^2+1)(x-1)}.$$

$$\text{Let } \frac{1}{x^3(x+1)^2(x^2+1)(x-1)} =$$

$$\frac{A+Bx}{(x^2+1)} + \frac{C}{(x+1)^2} + \frac{D}{(x+1)} + \frac{E}{(x-1)} + \frac{P}{x^3},$$

$$\therefore 1 = (A + Bv)(x-1)(x+1)^2 x^3 +$$

$$C(x^2+1)(x-1)x^3 + D(x^2+1)(x+1)(x-1)x^3 +$$

$$E(x^2+1)(x+1)^2 x^3 + P(x^2+1)(x+1)^2(x-1).$$

Let $x = \sqrt{-1}$,

$$\therefore 1 = (A+B\sqrt{-1})(1+\sqrt{-1})^2(\sqrt{-1}-1) \times -\sqrt{-1} =$$

$$(A + B\sqrt{-1})(2\sqrt{-1}-2) =$$

$$2A\sqrt{-1} - 2A - 2\sqrt{-1}B - 2B,$$

$\therefore$ by equating $2A\sqrt{-1} = 2B\sqrt{-1}, \quad \therefore A = B,$

$$2A + 2B = -1, \qquad \therefore A = B = -\tfrac{1}{4},$$

$$\therefore 1 + \frac{1}{4}(x+1)^3(x-1)x^3 = \frac{1}{4}(x^7 + 2x^6 - 2x^4 - x^3 + 4)$$

$$= \frac{1}{4}(x^2+1)(x^5 + 2x^4 - x^3 - 4x^2 + 4)$$

$$= (x^2+1)\{C(x-1)x^3 + D(x+1)(x-1)x^3$$

$$+ E(x+1)^2 x^3 + P(x+1)^2(x-1)\},$$

$$\therefore \frac{1}{4}(x^5 + 2x^4 - x^3 - 4x^2 + 4) =$$

$$C(x-1)x^3 + D(x+1)(x-1)x^3 + E(x+1)^2 x^3$$

$$+ P(x+1)^2(x-1).$$

Let $x = -1, \quad \therefore \frac{1}{2} = 2C, \quad \therefore C = \frac{1}{4},$

$$\therefore \frac{1}{4}(x^5 + 2x^4 - x^3 - 4x^2 + 4 - x^4 + x^3)$$

$$= \frac{1}{4}(x^5 + x^4 - 4x^2 + 4) = \frac{1}{4}(x+1)(x^4 - 4x + 4)$$

$$= (x+1)\{D(x-1)x^3 + E(x+1)x^3 + P(x+1)(x-1)\},$$

$$\therefore \frac{1}{4}(x^4 - 4x + 4)$$

$$= D(x-1)x^3 + E(x+1)x^3 + P(x+1)(x-1).$$

Let $x = -1$, $\qquad \therefore \frac{9}{4} = 2D$, $\qquad \therefore D = \frac{9}{8}$,

$$\therefore \frac{1}{4}(x^4 - 4x + 4) - \frac{9}{8}(x^4 - x^3) = -\frac{1}{8}(7x^4 - 9x^3 + 8x - 8)$$

$$= -\frac{1}{8}(x+1)(7x^3 - 16x^2 + 16x - 8)$$

$$= (x+1)\{Ex^3 + P(x-1)\},$$

$$\therefore -\frac{1}{8}(7x^3 - 16x^2 + 16x - 8) = Ex^3 + P(x-1).$$

Let $x = 1$, $\qquad \therefore E = \frac{1}{8}$,

$$\therefore -\frac{1}{8}(8x^3 - 16x^2 + 16x - 8) = P(x-1),$$

$$\therefore -(x-1)(x^2 - x + 1) = P(x-1),$$

$$\therefore P = -(x^2 - x + 1),$$

$$\therefore \frac{1}{x^8 + x^7 - x^4 - x^3} = -\frac{1}{4}\frac{1+x}{1+x^2}$$

$$+ \frac{1}{4} \cdot \frac{1}{(1+x)^2} + \frac{9}{8} \cdot \frac{1}{1+x} + \frac{1}{8} \cdot \frac{1}{x-1} - \frac{x^2 - x + 1}{x^3},$$

$$\therefore \int \frac{dx}{x^5 + x^7 - x^4 - x^3} = -\frac{1}{4}\int \frac{1+x}{1+x^2}dx$$

$$+\frac{1}{4}\int \frac{dx}{(1+x)^2} + \frac{9}{8}\int \frac{dx}{1+x}$$

$$+\frac{1}{8}\int \frac{dx}{x-1} - \int \frac{dx}{x} + \int \frac{dx}{x^2} - \int \frac{dx}{x^1}$$

$$= -\frac{1}{8}\int \frac{2x\,dx}{1+x^2} - \frac{1}{4}\int \frac{dx}{x^2+1} + \frac{1}{4}\int \frac{dx}{x+1)^2}$$

$$+\frac{1}{8}\log(x+1)^9 + \frac{1}{8}\log(x-1) - \log x - \frac{1}{x} + \frac{1}{2\,x^2}$$

$$= \frac{1}{2\,x^2} - \frac{1}{x} - \frac{1}{8}\log(1+x^2) - \frac{1}{4}\tan^{-1}x - \frac{1}{4}\cdot\frac{1}{(x+1)}$$

$$+\frac{1}{8}\{\log(x+1)^9 + \log(x-1) - \log x^8\}$$

$$= \frac{2+2x-4x^2-4x-x^3}{4x^2(1+x)}$$

$$+\frac{1}{8}\{\log(x+1)^9 + \log(x-1) - \log(1+x^2) - \log x^8\}$$

$$-\frac{1}{4}\tan^{-1}x = \frac{2-2x-5x^2}{4x^2(1+x)}$$

$$+\log \sqrt[8]{\frac{(x+1)^9(x-1)}{(1+x^2)x^8}} - \frac{1}{4}\tan^{-1}x.$$

(0.) $\displaystyle\int \frac{(5x-2)\,dx}{x^3 + 6x^2 + 8x}$

$$\frac{5x-2}{x^1 + 6x^2 + 8x} = \frac{A}{x} + \frac{B}{x+2} + \frac{C}{x+4}$$

$$5x - 2 = A(x+2)(x+4) + Bx(x+4) + Cx(x+2).$$

$$\text{Let } x = \quad 0 \quad - 2 = \quad 8\,A, \therefore A = -\frac{1}{4}$$

$$x = -2 \quad -12 = -4\,B, \therefore B = \quad 8$$

$$x = -4 \quad -22 = \quad 8\,C, \therefore C = -\frac{11}{4},$$

$$\therefore \int \frac{(5x - 2)\,dx}{x^3 + 6x^2 + 8x}$$

$$= -\frac{1}{4}\int \frac{dx}{x} + 3\int \frac{dx}{x+2} - \frac{11}{4}\int \frac{dx}{x+4}$$

$$= -\frac{1}{4}\log x + \frac{12}{4}\log(x+2) - \frac{11}{4}\log(x+4)$$

$$= \frac{1}{4}\log \frac{(x+2)^{12}}{x(x+4)^{11}}.$$

(10.) $\int \dfrac{(3x + 1)\,dx}{x^3 + 2x^2 + x}$

$$\frac{3x + 1}{x^3 + 2x^2 + x} = \frac{A}{x} + \frac{B}{(x+1)^2} + \frac{C}{x+1},$$

$$3x + 1 = A(x+1)^2 + Bx + Cx(x+1).$$

Let $x = 0, \quad 1 = A,$

$$3x + 1 - x^2 - 2x - 1 = x(1 - x) = Bx + Cx(x+1),$$

$$\therefore 1 - x = B + C(x+1).$$

Let $x = -1, \quad 2 = B,$

$$\therefore -(1 + x) = C(x+1), \qquad \therefore C = -1,$$

$$\therefore \int \frac{(3x + 1)\,dx}{x^3 + 2x^2 + x} = \int \frac{dx}{x} + 2\int \frac{dx}{(x+1)^2} - \int \frac{dx}{x+1}$$

$$= \log x - \frac{2}{x+1} - \log(x+1)$$

$$= \log \frac{x}{x+1} - \frac{2}{x+1}.$$

(11.) $\displaystyle\int \frac{dx}{1+3x+2x^2}$

Let $\dfrac{1}{1+3x+2x^2} = \dfrac{A}{x+1} + \dfrac{B}{2x+1}$,

$$1 = A(2x+1) + B(x+1).$$

Let $x = -1$ $\qquad 1 = -A$

$x = -\dfrac{1}{2}$ $\qquad 1 = +\dfrac{B}{2}$ $\qquad B = 2,$

$$\therefore \int \frac{dx}{1+3x+2x^2} = -\int \frac{dx}{x+1} + \int \frac{2\,dx}{2x+1}$$

$$= -\log(x+1) + \log(2x+1) = \log\frac{2x+1}{x+1}.$$

(12.) $\displaystyle\int \frac{x\,dx}{(x-2)(x+3)^2}$

Let $\dfrac{x}{(x-2)(x+3)^2} = \dfrac{A}{(x+3)^2} + \dfrac{B}{x+3} + \dfrac{C}{x-2}$,

$$x = A(x-2) + (x+3)\{B(x-2) + C(x+3)\}.$$

Let $x = -3,$ $\qquad \therefore -3 = -5A,$ $\quad A = \dfrac{3}{5},$

$$\therefore \frac{5x - 3x + 6}{5} = \frac{2(x+3)}{5}$$

$$= (x+3)\{B(x-2) + C(x+3)\}$$

$$\therefore \frac{2}{5} = B(x-2) + C(x+3).$$

Let $x = -3$; $\quad \dfrac{2}{5} = -5B$, $\therefore B = -\dfrac{2}{25}$,

$\quad x = 2$; $\quad \dfrac{2}{5} = 5C$, $\therefore C = \dfrac{2}{25}$,

$$\therefore \int \frac{x\,dx}{(x-2)(x+3)^2}$$

$$= \frac{3}{5} \int \frac{dx}{(x+3)^2} - \frac{2}{25} \int \frac{dx}{x+3} + \frac{2}{25} \int \frac{dx}{x-2}$$

$$= -\frac{3}{5} \frac{1}{x+3} - \frac{2}{25} \log(x+3) + \frac{2}{25} \log(x-2)$$

$$= \frac{2}{25} \log \frac{x-2}{x+3} - \frac{3}{5} \frac{1}{x+3}.$$

(13.) $\displaystyle\int \frac{(x^2 + 3x + 1)\,dx}{x^3 + x^2 - 2x}$

Let $\dfrac{x^2 + 3x + 1}{x^3 + x^2 - 2x} = \dfrac{A}{x} + \dfrac{B}{x-1} + \dfrac{C}{x+2}$,

$$x^2 + 3x + 1 = A(x-1)(x+2) + Bx(x+2) + Cx(x-1)$$

Let $x = \quad 0 \qquad 1 = -2A,$ $\quad \therefore A = -\dfrac{1}{2}$

$\qquad x = \quad 1 \qquad 5 = \quad 3B,$ $\quad \therefore B = \quad \dfrac{5}{3}$

$\qquad x = -2 \quad -1 = \quad 6C,$ $\quad \therefore C = -\dfrac{1}{6}$,

$$\therefore \int \frac{(x^2 + 3x + 1)\,dx}{x^3 + x^2 - 2x}$$

$$= -\frac{1}{2} \int \frac{dx}{x} + \frac{5}{3} \int \frac{dx}{x-1} - \frac{1}{6} \int \frac{dx}{x+2}$$

$$= -\frac{1}{2} \log x + \frac{5}{3} \log (x-1) - \frac{1}{6} \log (x+2)$$

$$= \frac{1}{3} \log \frac{(x-1)^5}{\sqrt{x^1 + 2x^1}}$$

(14.) $\displaystyle\int \frac{x\,dx}{(x+1)(x+2)(x^2+1)}$

Let $\displaystyle\frac{x}{(x+1)(x+2)(x^2+1)} = \frac{A}{x+1} + \frac{B}{x+2} + \frac{P}{x^2+1}$

$x = A(x+2)(x^2+1) + B(x+1)(x^2+1) + P(x+1)(x+2).$

Let $x = -1;$ $\quad -1 = 2A,$ $\quad \therefore A = -\dfrac{1}{2}$

$\quad x = -2;$ $\quad -2 = -5B,$ $\quad \therefore B = +\dfrac{2}{5}$

$$x + \frac{x^3 + 2x^2 + x + 2}{2} - \frac{2x^3 + 2x^2 + 2x + 2}{5}$$

$$= \frac{x^3 + 6x^2 + 11x + 6}{10} = P(x+1)(x+2),$$

$$\therefore P = \frac{x+3}{10},$$

$$\therefore \int \frac{x\,dx}{(x+1)(x+2)(x^2+1)}$$

$$= -\frac{1}{2}\int \frac{dx}{x+1} + \frac{2}{5}\int \frac{dx}{x+2} + \frac{1}{10}\int \frac{x+3}{x^2+1}\,dx$$

$$= -\frac{1}{2}\log(x+1) + \frac{2}{5}\log(x+2) + \frac{1}{20}\log(x^2+1)$$

$$+ \frac{3}{10}\tan^{-1}x,$$

$$= \log \frac{(x^2 + 1)^{\frac{1}{10}} (x + 2)^{\frac{2}{3}}}{\sqrt{x + 1}} + \frac{3}{10} \tan^{-1} x.$$

(15.) $\displaystyle\int \frac{dx}{x^4 + 4x + 3}$

Let $\dfrac{1}{x^4 + 4x + 3} = \dfrac{A}{(x + 1)^2} + \dfrac{B}{x + 1} + \dfrac{Mx + N}{x^2 - 2x + 3}$

$$1 = A (x^2 - 2x + 3) + B(x + 1) (x^2 - 2x + 3)$$
$$+ (Mx + N) (x + 1)^2.$$

Let $x = -1$; $1 = 6A$ $\therefore A = \dfrac{1}{6}$

$$\frac{6 - x^2 + 2x - 3}{6} = - \frac{(x^2 - 2x - 3)}{6} = - \frac{(x + 1)(x - 3)}{6}$$

$$= B(x + 1) (x^2 - 2x + 3) + (Mx + N) (x + 1)^2,$$

$$\therefore \; - \frac{x - 3}{6} = B(x^2 - 2x + 3) + (Mx + N) (x + 1).$$

Let $x = -1$; $\dfrac{2}{3} = 6B$ $\therefore B = \dfrac{1}{9}$,

$$\therefore \; \frac{-9x + 27 - 6x^2 + 12x - 18}{54}$$

$$= \frac{-(2x^2 - x - 3)}{18} = (Mx + N) (x + 1)$$

$$Mx + N = \frac{3 - 2x}{18},$$

$$\therefore \int \frac{dx}{x^4 + 4x + 3} =$$

$$\frac{1}{6} \int \frac{dx}{(x+1)} = +\frac{1}{9} \int \frac{dx}{x+1} + \frac{1}{6} \int \frac{dx}{x^2 - 2x + 3}$$

$$-\frac{1}{9} \int \frac{x\,dx}{x^2 - 2x + 3}$$

$$= -\frac{1}{6} \frac{1}{x+1} + \frac{1}{9} \log (x+1) - \frac{1}{18} \log (x^2 - 2x + 3)$$

$$+ \left(\frac{1}{6} - \frac{1}{9} \right) \int \frac{dx}{x^2 - 2x + 3}$$

$$= -\frac{1}{6} \frac{1}{x+1} + \frac{1}{9} \log \frac{x+1}{\sqrt{x^2 - 2x + 3}}$$

$$+ \frac{1}{18\sqrt{2}} \tan^{-1} \frac{x-1}{\sqrt{2}}.$$

(16.) $\int \dfrac{dx}{x(1+x^3)} = \left(\text{putting} \dfrac{1}{z} = x \right) - \int \dfrac{z^4}{z^3 + 1} \cdot \dfrac{dz}{z^2}$

$$= -\frac{1}{3} \int \frac{3z^2\,dz}{z^3 + 1} = -\frac{1}{3} \log (z^3 + 1) = \frac{1}{3} \log \frac{x^3}{x^3 + 1}$$

$$= \log \sqrt[3]{\frac{x^3}{x^3 + 1}}.$$

(17.) $\int \dfrac{dx}{x(1+x^4)} = \left(\text{putting} \dfrac{1}{z} = x \right) - \int \dfrac{z^3 dz}{z^4 + 1} \cdot \dfrac{1}{z^2}$

$$= -\int \frac{z^3\,dz}{z^4 + 1} = -\frac{1}{4} \log (z^4 + 1) = \log \sqrt[4]{\frac{x^4}{x^4 + 1}}.$$

(18.) $\int \dfrac{dx}{x^4(a+bx^3)} = \left(\text{ if } \dfrac{1}{z} = x \right) - \int \dfrac{z^7}{az^3 + b} \cdot \dfrac{dz}{z^2}$

$$= -\int \dfrac{z^5\, dz}{az^3 + b} = -\int \dfrac{z^2\, dz}{a} - \dfrac{b}{a^2}\int \dfrac{az^2\, dz}{az^3 + b}$$

$$= -\dfrac{z^3}{3a} + \dfrac{b}{3a^2}\log\,(az^3 + b)$$

$$= -\dfrac{1}{3ax^3} + \dfrac{b}{3a^2}\log\left(\dfrac{a+bx^3}{x^3}\right).$$

(19.) $\int \dfrac{dx}{x\,(1 + x^3)^2}$.   Let $x = \dfrac{1}{z}$, $dx = -\dfrac{1}{z^2}dz$

$$\therefore \int \dfrac{dx}{x\,(1 + x^3)^2} = -\int \dfrac{z^7}{(z^3 + 1)^2}\dfrac{dz}{z^2} = -\int \dfrac{z^5\, dz}{(z^3 + 1)^2}$$

$$= -\int \dfrac{z^2\, dz}{(z^3 + 1)} + \int \dfrac{z^2\, dz}{(z^3 + 1)^2}$$

$$= -\dfrac{1}{3}\,\log\,(z^3 + 1) - \dfrac{1}{3\,(z^3 + 1)}$$

$$= -\sqrt[3]{\dfrac{x^3 + 1}{x^3}} - \dfrac{x^3}{3(x^3+1)}.$$

The following method of doing the last four integrals is very simple, and can be often used with advantage.

$$\int \dfrac{dx}{x\,(x^3 + 1)}$$

$$\dfrac{1}{x\,(x^3 + 1)} = \dfrac{(1 + x^3) - x^3}{x\,(x^3 + 1)}$$

$$= \dfrac{1}{x} - \dfrac{x^2}{x^3 + 1},$$

$$\therefore \int \dfrac{dx}{x\,(x^3 + 1)} = \log x - \dfrac{1}{3}\log\,(x^3 + 1)$$

$$= \log \sqrt[3]{\dfrac{x^3}{x^3 + 1}}.$$

And $\int \dfrac{dx}{x(1+x^4)}$ may be found in a similar manner

$$\frac{1}{x(1+x^4)} = \frac{(1+x^4)-x^4}{x(1+x^4)} = \frac{1}{x} - \frac{x^3}{(1+x^4)},$$

$$\therefore \int \frac{dx}{x(1+x^4)} = \log \sqrt[4]{\frac{x^4}{x^4+1}}$$

$$\int \frac{dx}{x^4(a+bx^3)}.$$

Here $\dfrac{1}{x^4(a+bx^3)} = \dfrac{1}{a}\left(\dfrac{a+bx^3-bx^3}{x^4(a+bx^3)}\right)$

$$= \frac{1}{ax^4} - \frac{b}{a} \cdot \frac{1}{x(a+bx^3)}$$

$$\frac{1}{x(a+bx^3)} = \frac{1}{a} \cdot \frac{a+bx^3-bx^3}{x(a+bx^3)}$$

$$= \frac{1}{ax} - \frac{bx^2}{a(a+bx^3)}.$$

Therefore, the original quantity is reduced to

$$\frac{1}{ax^4} - \frac{b}{a^2x} + \frac{b^2}{a^2} \cdot \frac{x^2}{(a+bx^3)},$$

and its integral is $- \dfrac{1}{3ax^3} + \dfrac{b}{3a^2}\log\left(\dfrac{a+bx}{x^3}\right)$

$$\int \frac{dx}{x(1+x^3)^4}.$$

$$\frac{1}{x(1+x^3)^2} = \frac{(1+x^3)-x^3}{x(1+x^3)^2}$$

$$= \frac{1}{x(1+x^3)} - \frac{x^3}{(1+x^3)^2}$$

$$\frac{1}{x(1+x^3)} = \frac{1+x^3-x^3}{x(1+x^3)} = \frac{1}{x} - \frac{x^3}{x^3+1}.$$

Therefore $\dfrac{1}{x\,(1+x^i)^2}$

$$= \frac{1}{x} - \frac{x^2}{x^3+1} - \frac{x^2}{(x^3+1)^2};$$

and the integral required is therefore

$$\log \sqrt[3]{\frac{x^3}{x^3+1}} + \frac{1}{3\,(x^2+1)}.$$

(20.) $\displaystyle\int \frac{dx}{x^3-1}$.

$$\int \frac{dx}{x^6-1} = \frac{1}{2}\int \left\{ \frac{dx}{x^3-1} - \frac{dx}{x^3+1} \right\},$$

$$\int \frac{dx}{x^3-1} = \int \frac{dx}{(x-1)\,(x^2+x+1)}.$$

Let $\dfrac{1}{x^3-1} = \dfrac{A}{x-1} + \dfrac{Bx+C}{x^2+x+1}$,

$$1 = A\,(x^2+x+1) + Bx + C\,(x-1).$$

Let $x^2+x+1=0$, or $x = \dfrac{\sqrt{-3}-1}{2}$.

$$1 = \left(\frac{B\sqrt{-3}-B}{2} + C \right) \left(\frac{\sqrt{-3}-3}{2} \right),$$

$$\therefore 4 = (2\,C - 4\,B)\sqrt{-3} - 6\,C,$$

$$\therefore C = -\frac{2}{3}, \quad B = -\frac{1}{3}.$$

Let $x=1$, $\therefore 1 = 3\,A$, $A = \dfrac{1}{3}$,

∴ the integral is reduced to

$$\frac{1}{3}\int\left\{\frac{dx}{x-1}-\frac{(x+2)\,dx}{x^2+x+1}\right\}$$

$$\int\frac{dx}{x-1}=\log(x-1).$$

$$\int\frac{(x+2)\,dx}{x^2+x+1}=\left(\text{if } x=z-\frac{1}{2}\right)\int\frac{\left(z-\frac{1}{2}+2\right)dz}{\left(z^2+\frac{3}{4}\right)}$$

$$=\int\frac{\left(z+\frac{3}{2}\right)dz}{z^2+\frac{3}{4}}$$

$$=\log\sqrt{z^2+\frac{3}{4}}+\frac{3}{\sqrt{3}}\tan^{-1}\frac{2z}{\sqrt{3}}$$

$$=\log\sqrt{x^2+x+1}+\frac{3}{\sqrt{3}}\tan^{-1}\left(\frac{2x-1}{\sqrt{3}}\right).$$

$$\therefore\int\frac{dx}{x^3-1}=\frac{1}{3}\log(x-1)-\frac{1}{3}\log\sqrt{x^2+x+1}$$

$$-\frac{1}{\sqrt{3}}\tan^{-1}\left(\frac{2x-1}{\sqrt{3}}\right).$$

And in a similar manner it may be shown that

$$\int\frac{dx}{x^3+1}=\frac{1}{3}\log(x+1)-\frac{1}{3}\log\sqrt{x^2-x+1}$$

$$+\frac{1}{\sqrt{3}}\tan^{-1}\left(\frac{2x+1}{\sqrt{3}}\right),$$

$$\therefore\int\frac{dx}{x^6-1}=\frac{1}{6}\log\left(\frac{x-1}{x+1}\frac{\sqrt{x^2-x+1}}{\sqrt{x^2+x+1}}\right)$$

$$-\frac{1}{2\sqrt{3}}\left\{\tan^{-1}\left(\frac{2x-1}{\sqrt{3}}\right)+\tan^{-1}\left(\frac{2x+1}{\sqrt{3}}\right)\right\}.$$

But these inverse tangents together are equal to

$$\tan^{-1}\left(\frac{\dfrac{4x}{\sqrt{3}}}{1-\left(\dfrac{4x^2-1}{3}\right)}\right) = \tan^{-1}\left(\frac{x\sqrt{3}}{1-x^2}\right)$$

$$\therefore \int \frac{dx}{x^6-1} = \frac{1}{6}\log\left(\frac{x-1}{x+1}\frac{\sqrt{x^2-x+1}}{\sqrt{x^2+x+1}}\right)$$

$$-\frac{1}{2\sqrt{3}}\tan^{-1}\left(\frac{x\sqrt{3}}{1-x^2}\right),$$

(21.) $\displaystyle\int \frac{x^5\,dx}{(x^2+1)^3}$

$$\frac{x^5}{(x^2+1)^3} = \frac{(1+x^2)x^3-x^3}{(x^2+1)^3}$$

$$= \frac{x^3}{(x^2+1)^2} - \frac{x^3}{(x^2+1)^3}.$$

And $\quad \dfrac{x^3}{(x^2+1)^2} = \dfrac{x(x^2+1)-x}{(x^2+1)^2}$

$$= \frac{x}{x^2+1} - \frac{x}{(x^2+1)^2}.$$

Also $\quad \dfrac{x^3}{(1+x^2)^3} = \dfrac{x(x^2+1)-x}{(x^2+1)^3}$

$$= \frac{x}{(x^2+1)^2} - \frac{x}{(x^2+1)^3},$$

$$\therefore \frac{x^5}{(x^2+1)^3} = \frac{x^3}{(x^2+1)^2} - \frac{x^3}{(x^2+1)^3}$$

$$= \frac{x}{x^2+1} - \frac{2x}{(x^2+1)^2} + \frac{x}{(x^2+1)^3};$$

and the integral required is therefore

$$\log \sqrt{x^2 + 1} \;+\; \frac{1}{x^2 + 1} \;-\; \frac{1}{4(x^2 + 1)^2}$$

$$= \frac{4x^2 + 3}{4(x^2 + 1)^2} + \log \sqrt{x^2 + 1}.$$

(22.) $\displaystyle\int \frac{x^2\, dx}{x^4 + 1}$,

$$(x^m + 1) = \left(x^2 - 2x \, \cos \frac{\pi}{m} + 1 \right) \left(x^2 - 2x \, \cos \frac{3\pi}{m} + 1 \right) \cdots$$

continued to the factor $\left(x^2 - 2x \, \cos \dfrac{m-1}{m}\pi + 1 \right)$ when m is an even number.

This gives

$$(x^4 + 1) = \left(x^2 - 2x \, \cos \frac{\pi}{4} + 1 \right) \left(x^2 - 2x \, \cos \frac{3\pi}{4} + 1 \right)$$

or $(x^4 + 1) = (x^2 - x\sqrt{2} + 1)(x^2 + x\sqrt{2} + 1).$

Since, $\dfrac{\pi}{4} = 45^{\circ} \cos \dfrac{\pi}{4} = \sin \dfrac{\pi}{4} = \dfrac{1}{\sqrt{2}}, \ \cos \dfrac{3\pi}{4} = -\sin \dfrac{\pi}{4}.$

Assume therefore

$$\frac{x^2}{x^4 + 1} = \frac{Ax + B}{x^2 - x\sqrt{2} + 1} + \frac{Cx + D}{x^2 + x\sqrt{2} + 1};$$

$$\therefore x^2 = (Ax + B)(x^2 + x\sqrt{2} + 1) + (Cx + D)(x^2 - x\sqrt{2} + 1$$

If $x^2 + x\sqrt{2} + 1 = 0, \quad x = \dfrac{\sqrt{-1} - 1}{\sqrt{2}}$ (1.)

If $x^2 - x\sqrt{2} + 1 = 0, \quad x = \dfrac{\sqrt{-1} + 1}{\sqrt{2}}$ (2.)

Making (1) our supposition, we have

$$x^2 = -\frac{2\sqrt{-1}}{2} = -\sqrt{-1} = \frac{C\sqrt{-1}-C}{\sqrt{2}} + D.(2-2\sqrt{-1}),$$

or, $-\sqrt{-1} = \frac{2C\sqrt{-1}}{\sqrt{2}} + \frac{2C\sqrt{-1}}{\sqrt{2}} - \frac{2C}{\sqrt{2}} + \frac{2C}{\sqrt{2}}$

$$+ 2D - 2D\sqrt{-1}\cdot\sqrt{-1} = \left\{\frac{4C}{\sqrt{2}} - 2D\right\}\sqrt{-1} + 2D,$$

$$\therefore D = 0 \quad C = -\frac{1}{2\sqrt{2}}.$$

Now making (2) our supposition, we have

$$x^2 = \sqrt{-1} = \left\{\left(\frac{A\sqrt{-1}+A}{\sqrt{2}}\right) + B\right\}\left(2 + 2\sqrt{-1}\right),$$

= (by reduction, as in the preceding case,)

$$\left\{\frac{4A}{\sqrt{2}} + 2B\right\}\sqrt{-1} + 2B, \quad \therefore B = 0, \ A = \frac{1}{2\sqrt{2}}.$$

The quantity under consideration is now reduced to

$$\frac{1}{2\sqrt{2}}\left\{\frac{x}{x^2 - x\sqrt{2}+1} - \frac{x}{x^2 + x\sqrt{2}+1}\right\}.$$

The integrals of $\dfrac{x}{x^2-x\sqrt{2}+1}$ and $\dfrac{x}{x^2+x\sqrt{2}+1}$ now re-

main to be found. They can both be included in the general

case $\dfrac{x}{x^2 \pm x\sqrt{2}+1}$,

$$\int\frac{x\,dx}{x^2 \pm x\sqrt{2}+1} = \left(\text{if } x = z \mp \frac{1}{\sqrt{2}}\right)\int\frac{\left(z \mp \frac{1}{\sqrt{2}}\right)dz}{z^2 + \frac{1}{2}}$$

$$= \log\sqrt{z^2 + \frac{1}{2}}$$

$$\mp \tan^{-1}(z\sqrt{2}). \quad \text{This gives} \int \frac{x\,dx}{x^2 - x\sqrt{2}+1}$$

$$= \log\sqrt{(x^2 - x\sqrt{2}+1)} + \tan^{-1}(x\sqrt{2}-1),$$

$$\int \frac{x\,dx}{x^2 + x\sqrt{2}+1} = \log\sqrt{(x^2 + x\sqrt{2}+1)}$$

$$- \tan^{-1}(x\sqrt{2}+1), \qquad \therefore 2\sqrt{2}\int_x \frac{x^2}{x^4+1} =$$

$$\log\sqrt{\frac{x^2 - x\sqrt{2}+1}{x^2 + x\sqrt{2}+1}} + \tan^{-1}(x\sqrt{2}-1)$$

$$+ \tan^{-1}(x\sqrt{2}+1).$$

And these inverse tangents may be reduced thus:

$$\tan^{-1}(x\sqrt{2}-1) + \tan^{-1}(x\sqrt{2}+1)$$

$$= \tan^{-1}\left(\frac{x\sqrt{2}-1+x\sqrt{2}+1}{1-(x\sqrt{2}-1)(x\sqrt{2}+1)}\right)$$

$$= \tan^{-1}\frac{2x\sqrt{2}}{1-(2x^2-1)} = \tan^{-1}\frac{2x\sqrt{2}}{2(1-x^2)} = \tan^{-1}\frac{x\sqrt{2}}{(1-x^2)}.$$

$$\therefore \int \frac{x^2\,dx}{x^4+1} =$$

$$\frac{1}{2\sqrt{2}}\log\sqrt{\frac{x^2 - x\sqrt{2}+1}{x^2 + x\sqrt{2}+1}} + \frac{1}{2\sqrt{2}}\tan^{-1}\left\{\frac{x\sqrt{2}}{1-x^2}\right\}.$$

(23.) $\displaystyle\int \frac{x^6\,dx}{x^3 + 1}$

$$\frac{x^6}{x^3+1} = x^3 - 1 + \frac{1}{x^3+1}$$

$$\frac{1}{x^3+1} = \frac{A}{x+1} + \frac{Bx+C}{x^2-x+1},$$

c 3

$$\therefore\; 1 = A\,(x^2 - x + 1) + Bx + C\,(x + 1)$$

$$x = -1, \qquad \therefore\; 1 = 3A, \qquad \therefore\; A = \frac{1}{3},$$

$$\therefore\; \frac{3 - (x^2 - x + 1)}{x + 1} = -\frac{x^2 + x + 2}{x + 1}$$

$$= -\frac{x^2 - x + 2x + 2}{(x + 1)} = -x + 2 = Bx + C,$$

$$\therefore\; \frac{1}{x^3 + 1} = \frac{1}{3\,(x + 1)} + \frac{2 - x}{x^2 - x + 1}$$

$$\int \frac{(2 - x)\,dx}{x^2 - x + 1} = \left(\text{if } x = z + \frac{1}{2}\right) \int \frac{\left\{2 - \left(z + \frac{1}{2}\right)\right\}\,dz}{z^2 + \frac{3}{4}}$$

$$= \int \frac{\frac{3}{2}\,dz}{z^2 + \frac{3}{4}} - \int \frac{z\,dz}{z^2 + \frac{3}{4}}$$

$$= \frac{1}{\sqrt{3}}\tan^{-1}\left(\frac{2z}{\sqrt{3}}\right) - \log\sqrt{z^2 + \frac{3}{4}}$$

$$= \frac{1}{\sqrt{3}}\tan^{-1}\left(\frac{2x - 1}{\sqrt{3}}\right) - \log\sqrt{x^2 - x + 1},$$

$$\text{and } \int \frac{dx}{3\,(x + 1)} = \frac{1}{3}\log\,(x + 1),$$

$$\therefore\; \int \frac{dx}{x^3 + 1} = \frac{1}{\sqrt{3}}\tan^{-1}\left(\frac{2x - 1}{\sqrt{3}}\right) +$$

$$\log\frac{\sqrt[3]{x + 1}}{\sqrt{x^2 - x + 1}}.$$

Also, $\int x^3 dx = \dfrac{x^4}{4}, \quad \int dx = x,$

therefore the original integral is

$$\dfrac{x^4}{4} - x + \dfrac{1}{\sqrt{3}} \tan^{-1}\left\{\dfrac{2x-1}{\sqrt{3}}\right\} + \log \dfrac{\sqrt[3]{x+1}}{\sqrt{x^2-x+1}}$$

The method of integrating by parts is used to great advan tage in many integrals, which is as follows :—

$$d(pq) = p\,dq + q\,dp,$$

$$\therefore p\,dq = d(pq) - q\,dp$$

$$\int p\,dq = pq - \int q\,dp\ldots\ldots(1);$$

or by using differential coefficients

$$\int p\,\dfrac{dq}{dx} = pq - \int q\,\dfrac{dp}{dx}\ldots\ldots(2).$$

In the following examples we shall sometimes use (1) an sometimes (2).

When integrals are of the form of $\int x^{m-1}(a+bx^n)^{\frac{p}{q}}\,dx$

they can be rationalized by assuming $a+bx^n = z^q$ when

$\dfrac{m}{n}$ or $\dfrac{m}{n} + \dfrac{p}{q}$ is an integer. If $\dfrac{m}{n}$ be a fraction assume $a+bx$

$= x^n z^q.$

CHAPTER III.

(1.) $\int x\, dx \sqrt{a+x} = \int dx (a+x-a)\sqrt{a+x}$

$$= \int dx (a+x)^{\frac{3}{2}} - a \int dx \sqrt{a+x}$$

$$= \frac{2}{5}(a+x)^{\frac{5}{2}} - \frac{2a}{3}(a+x)^{\frac{3}{2}}.$$

(2.) $\dfrac{du}{dx} = \dfrac{1}{\sqrt{x+a}+\sqrt{x}} = \dfrac{\sqrt{x+a}-\sqrt{x}}{a}$,

$$\therefore u = \frac{2}{3a}(x+a)^{\frac{3}{2}} - \frac{2}{3a}x^{\frac{3}{2}}.$$

(3.) $\dfrac{du}{dx} = (a-x)(b-x)^{\frac{m}{n}} = (b-x+a-b)(b-x)^{\frac{m}{n}}$

$$= (b-x)^{\frac{m}{n}+1} + (a-b)(b-x)^{\frac{m}{n}},$$

$$\therefore u = -\frac{n(b-x)^{\frac{m}{n}+2}}{m+2n} - \frac{n(a-b)(b-x)^{\frac{m}{n}+1}}{m+n)}.$$

(4.) $\dfrac{d\,u}{d\,x} = (x^2 + a)\sqrt{x^2 + 4a}$

$$= (x^2 + 4a - 3a)\sqrt{x^2 + 4a}$$

$$= (x^2 + 4a)^{\frac{3}{2}} - 3a\sqrt{x^2 + 4a}.$$

To integrate $(x^2 + 4a)^{\frac{3}{2}}$ (integrating by parts).

Let $p = (x^2 + 4a)^{\frac{3}{2}}$ $\dfrac{d\,q}{d\,x} = 1$

$$\dfrac{dp}{dx} = 3x\,(x^2 + 4b)^{\frac{1}{2}}\qquad q = x,$$

$$\therefore \int dx\,(x^2 + 4a)^{\frac{3}{2}} = x\,(x^2 + 4a)^{\frac{3}{2}} - 3\int x^2\,dx\,(x^2 + 4a)^{\frac{1}{2}}$$

$$= x\,(x^2 + 4a)^{\frac{3}{2}} - 3\int dx\,(x^2 + 4a)^{\frac{3}{2}} + 12a\int d\,x\,(x^2 + 4a)^{\frac{1}{2}},$$

$$\therefore \int dx\,(x^2 + 4a)^{\frac{3}{2}} = \dfrac{x}{4}(x^2 + 4a)^{\frac{3}{2}} + 3a\int dx\,(x^2 + 4a)^{\frac{1}{2}},$$

$$\therefore u = \dfrac{x}{4}(x^2 + 4a)^{\frac{3}{2}}.$$

(5.) $\dfrac{d\,u}{d\,x} = \dfrac{1}{x\sqrt{x + a}}.$ Let $a + x = z^2$

$$\dfrac{d\,u}{d\,z} = \dfrac{d\,u}{d\,x}\dfrac{d\,x}{d\,z} = \dfrac{1}{(z^2 - a)z}\cdot 2z = \dfrac{2}{z^2 - a}$$

$$= \dfrac{1}{\sqrt{a}}\left\{\dfrac{1}{z - \sqrt{a}} - \dfrac{1}{z + \sqrt{a}}\right\}$$

$$u = \dfrac{1}{\sqrt{a}}\log\dfrac{z - \sqrt{a}}{z + \sqrt{a}} = \dfrac{1}{\sqrt{a}}\log\dfrac{z^2 - a}{(z + \sqrt{a})^2}$$

$$= \frac{2}{\sqrt{a}} \log \frac{\sqrt{x}}{\sqrt{a+x}+\sqrt{a}}.$$

(6.) $\dfrac{du}{dx} = \dfrac{\sqrt{a+bx^3}}{x}$. Let $a + bx^3 = z^2$,

$$x = \left(\frac{z^2-a}{b}\right)^{\frac{1}{3}}, \quad \frac{dx}{dz} = \frac{1}{b^{\frac{1}{3}}} \frac{2z}{3(z^2-a)^{\frac{2}{3}}}$$

$$\frac{du}{dz} = \frac{du}{dx} \cdot \frac{dx}{dz} = \frac{b^{\frac{1}{3}}z}{(z^2-a)^{\frac{1}{3}}} \cdot \frac{2z}{3b^{\frac{1}{3}}(z^2-a)^{\frac{2}{3}}}$$

$$= \frac{2z^2}{3(z^2-a)} = \frac{2}{3}\left\{1 + \frac{a}{z^2-a}\right\}$$

$$= \frac{2}{3}\left\{1 + \frac{\sqrt{a}}{2}\left(\frac{1}{z-\sqrt{a}} - \frac{1}{z+\sqrt{a}}\right)\right\},$$

$$\therefore u = \frac{2z}{3} - \frac{2}{3} \cdot \frac{\sqrt{a}}{2} \log \frac{z+\sqrt{a}}{z-\sqrt{a}}$$

$$= \frac{2z}{3} - \frac{2\sqrt{a}}{3} \log \frac{z+\sqrt{a}}{\sqrt{z^2-a}}$$

$$= \frac{2\sqrt{a+bx^3}}{3} - \frac{2\sqrt{a}}{3} \log \frac{\sqrt{a+bx^3}+\sqrt{a}}{\sqrt{bx^3}}$$

$$= \frac{2\sqrt{a+bx^3}}{3} - \frac{2\sqrt{a}}{3} \log \frac{\sqrt{b+ax^{-3}}+\sqrt{ax^{-3}}}{\sqrt{b}}.$$

(7.) $\dfrac{du}{dx} = \dfrac{1+x^2}{(1-x^2)\sqrt{1+x^4}} = \dfrac{\frac{1}{x^2}+1}{\left(\frac{1}{x}-x\right)\sqrt{\frac{1}{x^2}+x^2}}$

$$= \dfrac{\frac{1}{x^2}+1}{\left(\frac{1}{x}-x\right)\sqrt{\left(\frac{1}{x}-x\right)^2+2}},$$

$$= \dfrac{-dz}{z\sqrt{z^2+2}}\left(\text{if } \frac{1}{x}-x=z\right),$$

$$\therefore u = -\dfrac{1}{\sqrt{2}}\log\left\{\dfrac{z}{\sqrt{z^2+2}+\sqrt{2}}\right\}$$

$$= \dfrac{1}{\sqrt{2}}\log\left\{\dfrac{\sqrt{z^2+2}+\sqrt{2}}{z}\right\}$$

$$= \dfrac{1}{\sqrt{2}}\log\dfrac{\sqrt{2}x+\sqrt{1+x^4}}{1-x^2}.$$

(8.) $\displaystyle\int\dfrac{dx}{\sqrt{x^2+x+1}} = \int\dfrac{dx}{\sqrt{(x+\frac12)^2+\frac34}}$

$$= \log\left\{\sqrt{(x+\tfrac12)^2+\tfrac34}+x+\tfrac12\right\}$$

$$= \log\left\{2\sqrt{x^2+x+1}+2x+1\right\}+\text{c}.$$

(9.) $\displaystyle\int\dfrac{dx}{\sqrt{1+x-x^3}} = \int\dfrac{dx}{\sqrt{\frac53-(x-\frac13)^2}}$

$$= \sin^{-1} \frac{x - \frac{1}{2}}{\frac{\sqrt{5}}{2}} = \sin^{-1} \frac{2x - 1}{\sqrt{5}}.$$

(10.) $\int \dfrac{dx}{(1 + x)\sqrt{1 - x}}$ Let $1 - x = z^2,$

$$x = 1 - z^2 \qquad dx = -2z\,dz$$

$$1 + x = 2 - z^2$$

$$\int \frac{dx}{(1 + x)\sqrt{1 - x}} = -2 \int \frac{dz}{2 - z^2}$$

$$= -\frac{2}{2\sqrt{2}} \left\{ \int \frac{dz}{\sqrt{2} + z} + \int \frac{dz}{\sqrt{2} - z} \right\}$$

$$= -\frac{1}{\sqrt{2}} \log \frac{\sqrt{2} - z}{\sqrt{2} + z} = \frac{1}{\sqrt{2}} \log \frac{2 + z^2 - 2\sqrt{2}z}{2 - z^2}$$

$$= \frac{1}{\sqrt{2}} \log \frac{3 - x - 2\sqrt{2}\sqrt{1 - x}}{1 + x}.$$

(11.) $\int \dfrac{dx}{x\sqrt{1 + x + x^2}}$.   Let $x = \dfrac{1}{z}$ $dx = -\dfrac{dz}{z^2}$

$$= -\int \frac{dz}{\sqrt{1 + z + z^2}} = -\int \frac{dz}{\sqrt{(z + \frac{1}{2})^2 + \frac{3}{4}}}$$

$$= -\log \left\{ \sqrt{1 + z + z^2} + z + \tfrac{1}{2} \right\}$$

$$= -\log \left\{ \frac{2\sqrt{1 + x + x^2} + 2 + x}{x} \right\}$$

$$= \log \frac{x}{2\sqrt{1 + x + x^2} + 2 + x}$$

$$= \log \frac{x\left(2 + x - 2\sqrt{1 + x + x^2}\right)}{4 + 4x + x^2 - 4 - 4x - 4x^2}$$

$$= \log \frac{2 + x - 2\sqrt{1 + x + x^2}}{-3x}$$

(12.) $\dfrac{d\,u}{d\,x} = \dfrac{x}{\sqrt{a^4 - x^4}} = \dfrac{1}{2} \dfrac{\dfrac{d\,(x^2)}{d\,x}}{\sqrt{a^4 - (x^2)^2}},$

$$\therefore u = \frac{1}{2} \sin^{-1} \frac{x^2}{a^2}.$$

(13.) $\dfrac{d\,u}{d\,x} = \dfrac{1}{\sqrt{(1 - x)(x + 2)}} = \dfrac{1}{\sqrt{2 - x - x^2}}$

$$= \frac{2}{\sqrt{9 - (2x + 1)^2}},$$

$$\therefore u = \sin^{-1} \frac{2x + 1}{3}.$$

(14.) $\dfrac{d\,u}{d\,x} = \dfrac{x}{\sqrt{(x^2 - a^2)(b^2 - x^2)}}$

$$= \frac{x}{\sqrt{x^2 - a^2}} \frac{1}{\sqrt{b^2 - a^2 - (x^2 - a^2)}},$$

$$\therefore u = \sin^{-1} \sqrt{\frac{x^2 - a^2}{b^2 - a^2}}.$$

(15) $\dfrac{du}{dx} = \dfrac{1}{x\sqrt{x^2 - a^2}\sqrt{b^2 - x^2}}$

Let $x = \dfrac{1}{z}$ $\dfrac{dx}{dz} = -\dfrac{1}{z^2}$,

$$\frac{du}{dz} = \frac{du}{dx} \cdot \frac{dx}{dz} = \frac{z^3}{\sqrt{1 - a^2 z^2}\ \sqrt{b^2 z^2 - 1}} \cdot -\frac{1}{z^4}$$

$$= -\frac{a^2 z}{\sqrt{1 - a^2 z^2}}\ \frac{1}{a^2 \cdot \dfrac{b}{a}\sqrt{1 - \dfrac{a^2}{b^2} - (1 - a^2 z^2)}},$$

$$\therefore u = \frac{1}{ab}\ \sin^{-1}\left\{ \frac{1 - a^2 z^2}{1 - \dfrac{a^2}{b^2}} \right\}^{\frac{1}{2}}$$

$$= \frac{1}{ab}\ \sin^{-1}\frac{b}{x}\sqrt{\frac{x^2 - a^2}{b^2 - a^2}}.$$

(16.) $\dfrac{du}{dx} = \dfrac{1}{(1 + x)\sqrt{1 - x^2}}$

$$= \frac{1}{(1 + x)\sqrt{2(1 + x) - (1 + x)^2}}$$

$$= \frac{(1 + x)^{-2}}{\sqrt{2(1 + x)^{-1} - 1}}.$$

$$\therefore u = -\sqrt{2(1 + x)^{-1} - 1}$$

$$= -\sqrt{\frac{2}{1 + x} - 1} = -\sqrt{\frac{1 - x}{1 + x}}.$$

(17.) $\dfrac{du}{dx} = \dfrac{x}{(1 + x^2)\sqrt{1 - x^4}}$

$$= \frac{x}{(1 + x^2)\sqrt{2(1 + x^2) - (1 + x^2)^2}} = \frac{x(1 + x^2)^{-2}}{\sqrt{2(1 + x^2)^{-1} - 1}},$$

$$\therefore u = -\frac{1}{2}\sqrt{2(1 + x^2)^{-1} - 1} = -\frac{1}{2}\sqrt{\frac{1 - x^2}{1 + x^2}}.$$

(18.) $\dfrac{du}{dx} = \dfrac{(x + \sqrt{1 + x^2})^{\frac{m}{n}}}{\sqrt{1 + x^2}}$

$= \left(\dfrac{x}{\sqrt{1 + x^2}} + 1 \right) (x + \sqrt{1 + x^2})^{\frac{m}{n} - 1}$,

$\therefore\ u = \dfrac{n}{m} (x + \sqrt{1 + x^2})^{\frac{m}{n}}$.

(19.) $\dfrac{du}{dx} = \dfrac{1}{(1 + x)\sqrt{1 + x - x^2}}$.

Let $1 + x = z,$ $\quad x^2 = (z - 1)^2,$

$\dfrac{du}{dz} = \dfrac{du}{dx} \cdot \dfrac{dx}{dz} = \dfrac{1}{z\sqrt{z - (z - 1)^2}} = \dfrac{1}{z\sqrt{3z - z^2 - 1}}$;

Let $z = \dfrac{1}{v},$ $\dfrac{dz}{dv} = -\dfrac{1}{v^3}$

$\dfrac{du}{dv} = \dfrac{du}{dz} \cdot \dfrac{dz}{dv} = -\dfrac{1}{v^2} \dfrac{v^2}{\sqrt{3v - 1 - v^2}}$

$= \dfrac{-1}{\sqrt{\dfrac{5}{4} - \left(\dfrac{3}{2} - v\right)^2}}$,

$\therefore\ u = -\sin^{-1} \dfrac{\dfrac{3}{2} - v}{\dfrac{\sqrt{5}}{2}} = -\sin^{-1} \dfrac{3 - 2v}{\sqrt{5}}$

$= -\sin^{-1} \dfrac{3z - 2}{\sqrt{5}z} = -\sin^{-1} \dfrac{3x + 1}{\sqrt{5}(x + 1)}$.

$= \tan^{-1} \left(\dfrac{3x + 1}{2\sqrt{1 + x - x^2}} \right)$.

(20.) $\dfrac{du}{dx} = \dfrac{1}{(x+b)\sqrt{x+a}}.$

Let $x + a = z^2$ $\dfrac{dx}{dz} = 2z,$

$\dfrac{du}{dz} = \dfrac{du}{dx} \cdot \dfrac{dx}{dz} = \dfrac{1}{(z^2 - a + b)z} \cdot 2z = \dfrac{2}{z^2 + b - a},$

$\therefore\ u = \dfrac{2}{\sqrt{b-a}} \tan^{-1} \dfrac{z}{\sqrt{b-a}}$

$\qquad\qquad = \dfrac{2}{\sqrt{b-a}} \tan^{-1} \sqrt{\dfrac{x+a}{b-a}}.$

(21.) $\dfrac{du}{dx} = \dfrac{1}{(x+b)(x+a)^{\frac{3}{2}}}$

$\qquad = \dfrac{1}{a-b}\left\{\dfrac{1}{x+b} - \dfrac{1}{x+a}\right\}\dfrac{1}{\sqrt{x+a}},$

$\qquad = \dfrac{1}{a-b} \cdot \dfrac{1}{(x+b)\sqrt{x+a}} - \dfrac{1}{a-b} \cdot \dfrac{1}{(x+a)^{\frac{3}{2}}},$

$\therefore\ u = -\dfrac{2}{(b-a)^{\frac{3}{2}}} \tan^{-1} \sqrt{\dfrac{x+a}{b-a}} + \dfrac{2}{a-b}\dfrac{1}{\sqrt{x+a}}$

(22.) $\dfrac{du}{dx} = \dfrac{1}{x}\sqrt{x^2 - a^2} = \dfrac{1}{x}\dfrac{x^2 - a^2}{\sqrt{x^2 - a^2}}$

$\qquad = \dfrac{x}{\sqrt{x^2 - a^2}} - \dfrac{a^2}{x\sqrt{x^2 - a^2}},$

$\therefore\ u = \sqrt{x^2 - a^2} - a\sec^{-1}\dfrac{x}{a}.$

(23.) $\dfrac{du}{dx} = \dfrac{\sqrt{a^2 \pm x^2}}{x^2} = \dfrac{a^2}{x^2\sqrt{a^2 \pm x^2}} \pm \dfrac{1}{\sqrt{a^2 \pm x^2}}.$

To integrate $\dfrac{1}{x^2 \sqrt{a^2 \pm x^2}}.$

Let $x = \dfrac{1}{z}, \qquad \dfrac{dx}{dz} = -\dfrac{1}{z^2},$

$\dfrac{du}{dz} = \dfrac{dx}{dz} \cdot \dfrac{du}{dx} = -\dfrac{1}{z^2}\dfrac{z^3}{\sqrt{a^2 z^2 \pm 1}} = -\dfrac{z}{\sqrt{a^2 z^2 \pm 1}},$

$$\therefore \ u = -\dfrac{\sqrt{a^2 z^2 \pm 1}}{a^2},$$

$$\therefore \int \dfrac{\sqrt{a^2 \pm x^2}}{x^3} \, dx = -\dfrac{\sqrt{a^2 + x^2}}{x} + \log(x + \sqrt{a^2 + x^2}),$$

or, $= -\dfrac{\sqrt{a^2 - x^2}}{x} - \sin^{-1}\dfrac{x}{a},$ accordingly as we use $a^2 + x^2$ or $a^2 - x^2$.

(24.) $\dfrac{du}{dx} = \dfrac{1}{\sqrt{a^2 + x^2} - \sqrt{a^2 - x^2}}$

$$= \dfrac{\sqrt{a^2 + x^2} + \sqrt{a^2 - x^2}}{2x^2}$$

$$= \dfrac{1}{2}\dfrac{\sqrt{a^2 + x^2}}{x^2} + \dfrac{1}{2}\dfrac{\sqrt{a^2 - x^2}}{x^2}.$$

By last example,

$$u = -\dfrac{\sqrt{a^2 + x^2} + \sqrt{a^2 + x^2}}{2x}$$

$$+ \dfrac{1}{2}\log(x + \sqrt{a^2 + x^2}) - \dfrac{1}{2}\sin^{-1}\dfrac{x}{a}.$$

(25.) $\dfrac{du}{dx} = \sqrt{\dfrac{a+x}{a-x}} = \dfrac{a}{\sqrt{a^2-x^2}} + \dfrac{x}{\sqrt{a^2-x^2}}$,

$$\therefore\ u = a\sin^{-1}\dfrac{x}{a} - \sqrt{a^2-x^2}.$$

(26.) $\dfrac{du}{dx} = \dfrac{1}{x}\sqrt{\dfrac{x+a}{x-a}} = \dfrac{1}{\sqrt{x^2-a^2}} + \dfrac{a}{x\sqrt{x^2-a^2}}$,

$$\therefore\ u = \log\left(x + \sqrt{x^2-a^2}\right) + \sec^{-1}\dfrac{x}{a}.$$

(27.) $\dfrac{du}{dx} = x\sqrt{\dfrac{a+x}{a-x}} = \dfrac{ax}{\sqrt{a^2-x^2}} + \dfrac{x^2}{\sqrt{a^2-x^2}}.$

To integrate $\dfrac{x^2}{\sqrt{a^2-x^2}}$.

$$p = x \qquad \dfrac{dq}{dx} = \dfrac{x}{\sqrt{a^2-x^2}},$$

$$\dfrac{dp}{dx} = 1 \qquad q = -\sqrt{a^2-x^2}.$$

$$\therefore \int \dfrac{x^2\,dx}{\sqrt{a^2-x^2}} = -x\sqrt{a^2-x^2} + \int \sqrt{a^2-x^2}\,dx$$

$$= -x\sqrt{a^2-x^2} + \int \dfrac{a^2\,dx}{\sqrt{a^2-x^2}} - \int \dfrac{x^2\,dx}{\sqrt{a^2-x^2}}$$

$$\therefore \int \dfrac{x^2\,dx}{\sqrt{a^2-x^2}} = -\dfrac{x}{2}\sqrt{a^2-x^2} + \dfrac{a^2}{2}\int \dfrac{dx}{\sqrt{a^2-x^2}}.$$

$$\therefore \; u = a \int \frac{x\,dx}{\sqrt{a^2-x^2}} + \int \frac{x^2\,dx}{\sqrt{a^2-x^2}}$$

$$= \frac{a^2}{2}\sin^{-1}\frac{x}{a} - \left(a+\frac{x}{2}\right)\sqrt{a^2-x^2}.$$

(28.) $\quad \dfrac{du}{dx} = \dfrac{1}{\sqrt{1-2acx+a^2c^2}\,\sqrt{1-2ac^{-1}x+a^2c^{-2}}}.$

Let $1-2acx+a^2c^2 = z^2 \qquad \dfrac{dx}{dz} = -\dfrac{z}{ac}$

$$-2ac^{-1}x = \frac{z^2-(1+a^2c^2)}{c^2}$$

$$1-2ac^{-1}x+a^2c^{-2} = \frac{c^2+a^2+z^2-1-a^2c^2}{c^2},$$

$$\therefore \; \frac{du}{dz} = \frac{du}{dx}\cdot\frac{dx}{dz} = -\frac{z}{ac}\cdot\frac{c}{z\sqrt{z^2-(1-a^2-c^2+a^2c^2)}}$$

$$= -\frac{1}{a}\frac{1}{\sqrt{z^2-(1-a^2-c^2+a^2c^2)}},$$

$$\therefore \; u = -\frac{1}{a}\log\{z+\sqrt{z^2-(1-a^2-c^2+a^2c^2)}\}$$

$$= \frac{1}{a}\log\{\sqrt{1-2acx+a^2c^2}+\sqrt{a^2-2acx+c^2}\}$$

(29.) $\quad \dfrac{du}{dx} = \dfrac{1}{(1-x^2)\sqrt{1+x^2}}. \qquad$ Let $x = \dfrac{1}{z}$

$$\frac{du}{dz} = -\frac{1}{z^2}\frac{z^3}{(z^2-1)\sqrt{1+z^2}} = -\frac{z}{(z^2-1)\sqrt{1+z^2}}.$$

Let $1+z^2 = v^2, \quad z^2-1 = v^2-2, \quad \dfrac{dz}{dv} = \dfrac{v}{\sqrt{v^2-1}}$

$$\frac{du}{dv} = \frac{du}{dz} \cdot \frac{dz}{dv} = -\frac{\sqrt{v^2 - 1}}{(v^2 - 2)\, v} \cdot \frac{v}{\sqrt{v^2 - 1}} = -\frac{1}{v^2 - 2}$$

$$\frac{du}{dv} = \frac{1}{2\sqrt{2}}\left\{\frac{1}{v + \sqrt{2}} - \frac{1}{v - \sqrt{2}}\right\},$$

$$\therefore \; u = \frac{1}{2\sqrt{2}}\log\frac{v + \sqrt{2}}{v - \sqrt{2}} = \frac{1}{2\sqrt{2}}\log\frac{(v + \sqrt{2})^2}{v^2 - 2}$$

$$= \frac{1}{\sqrt{2}}\log\frac{\sqrt{1 + z^2} + \sqrt{2}}{\sqrt{z^2 - 1}}$$

$$= \frac{1}{\sqrt{2}}\log\frac{\sqrt{1 + x^2} + \sqrt{2}\,x}{\sqrt{1 - x^2}}$$

(30.) $\quad \dfrac{du}{dx} = \dfrac{1}{(1 + x^2)\sqrt{1 - x^2}}.$ Let $x = \dfrac{1}{z}$

$$\frac{du}{dz} = -\frac{z}{(1 + z^2)\sqrt{z^2 - 1}}.$$

Let $z^2 - 1 = v^2$, $\; 1 + z^2 = v^2 + 2$, $\; \dfrac{dz}{dv} = \dfrac{v}{\sqrt{v^2 + 1}}$,

$$\therefore \; \frac{du}{dv} = \frac{du}{dz} \cdot \frac{dz}{dv} = -\frac{\sqrt{v^2 + 1}}{(v^2 + 2)\, v} \cdot \frac{v}{\sqrt{v^2 + 1}} = -\frac{1}{v^2 + 2}$$

$$u = -\frac{1}{\sqrt{2}}\tan^{-1}\frac{v}{\sqrt{2}} = -\frac{1}{\sqrt{2}}\tan^{-1}\frac{\sqrt{z^2 - 1}}{\sqrt{2}}$$

$$= -\frac{1}{\sqrt{2}}\tan^{-1}\frac{\sqrt{1 - x^2}}{\sqrt{2}\,x}.$$

(31.) $\displaystyle\int \frac{dx}{x\sqrt{x^2+x-1}}$.　　Let $x=\dfrac{1}{z}, \quad dx=-\dfrac{dz}{z^2}$

$$=-\int \frac{dz}{\sqrt{1+z-z^2}}=-\int \frac{dz}{\sqrt{\dfrac{5}{4}-\left(z-\dfrac{1}{2}\right)^2}}$$

$$=-\sin^{-1}\frac{z-\dfrac{1}{2}}{\dfrac{\sqrt{5}}{2}}=-\sin^{-1}\frac{2z-1}{\sqrt{5}}$$

$$=-\sin^{-1}\frac{2-x}{\sqrt{5}\,x}=\sin^{-1}\frac{x-2}{\sqrt{5}\,x}.$$

(32.) $\displaystyle\int \frac{dx}{(1+x)\sqrt{1+x^2}}$.　　Let $1+x=\dfrac{1}{z}$

$$x=\frac{1}{z}-1, \qquad dx=-\frac{dz}{z^2}$$

$$1+x^2=2-\frac{2}{z}+\frac{1}{z^2},$$

$$\therefore \int \frac{dx}{(1+x)\sqrt{1+x^2}}=\int \frac{-\dfrac{dz}{z^2}}{\dfrac{1}{z}\sqrt{2-\dfrac{2}{z}+\dfrac{1}{z^2}}}$$

$$=-\int \frac{dz}{\sqrt{2z^2-2z+1}}=-\frac{1}{\sqrt{2}}\int \frac{dz}{\sqrt{\left(z-\dfrac{1}{2}\right)^2+\dfrac{1}{4}}}$$

$$=-\frac{1}{\sqrt{2}}\log\left\{\sqrt{z^2-z+\frac{1}{2}}+z-\frac{1}{2}\right\}$$

$$=-\frac{1}{\sqrt{2}}\log\left\{\frac{\sqrt{1+x^2}}{\sqrt{2}\,(1+x)}+\frac{1-x}{2\,(1+x)}\right\}$$

$$= -\frac{1}{\sqrt{2}}\log\left\{\frac{\sqrt{2}\sqrt{1+x^2}+1-x}{2(1+x)}\right\}$$

$$= \frac{1}{\sqrt{2}}\log\frac{2(1+x)}{1-x+\sqrt{2}\sqrt{1+x^2}}$$

$$= \frac{1}{\sqrt{2}}\log\frac{2(1+x)\{1-x-\sqrt{2}\sqrt{1+x^2}\}}{1-2x+x^2-2-2x^2}$$

$$= \frac{1}{\sqrt{2}}\log\left\{\frac{2(1-x-\sqrt{2}\sqrt{1+x^2})}{-(1+x)}\right\}.$$

(33.) $\int\frac{dx}{(1+x^2)\sqrt{1-x^2}}.$ Let $1-x^2 = x^2z^2$

$$x^2 = \frac{1}{z^2+1} \qquad 1+x^2 = \frac{2+z^2}{1+z^2}$$

$$2\log x = -\log(z^2+1),$$

$$\frac{dx}{x} = -\frac{zdz}{z^2+1}$$

$$\int\frac{dx}{(1+x^2)xz} = -\int\frac{dz}{z^2+1}\times\frac{z^2+1}{z^2+2} = -\int\frac{dz}{2+z^2}$$

$$= \frac{1}{\sqrt{2}}\cot^{-1}\frac{z}{\sqrt{2}} = \frac{1}{\sqrt{2}}\cot^{-1}\frac{\sqrt{1-x^2}}{\sqrt{2}x}.$$

Let $\theta = \cot^{-1}\frac{\sqrt{1-x^2}}{\sqrt{2}x}$

$$\cot\theta = \frac{\sqrt{1-x^2}}{\sqrt{2}x}$$

$$\cos\theta = \frac{\cot\theta}{\sqrt{1+\cot^2\theta}} = \frac{\frac{\sqrt{1-x^2}}{\sqrt{2}x}}{\sqrt{1+\frac{1-x^2}{2x^2}}}$$

$$\cos\theta = \sqrt{\frac{1-x^2}{1+x^2}}, \qquad \therefore \ \theta = \cos^{-1}\sqrt{\frac{1-x^2}{1+x^2}},$$

$$\therefore \int\frac{dx}{(1+x^2)\sqrt{1-x^2}} = \frac{1}{\sqrt{2}}\cos^{-1}\sqrt{\frac{1-x^2}{1+x^2}}.$$

(34.) $\displaystyle\int\frac{dx}{(1-x^2)\sqrt{(1+x^2)}}.$ Let $1+x^2 = x^2 z^2$

$$x^2 = \frac{1}{z^2-1} \qquad 1-x^2 = \frac{z^2-2}{z^2-1}$$

$$2\log x = -\log(z^2-1)$$

$$\frac{dx}{xz} = -\frac{dz}{z^2-1}$$

$$\int\frac{dx}{(1-x^2)xz} = -\int\frac{dz}{z^2-1}\times\frac{z^2-1}{z^2-2} = -\int\frac{dz}{z^2-2}$$

$$= \frac{1}{2\sqrt{2}}\left\{\int\frac{dz}{z+\sqrt{2}} - \int\frac{dz}{z-\sqrt{2}}\right\}$$

$$= \frac{1}{2\sqrt{2}}\log\frac{z+\sqrt{2}}{z-\sqrt{2}} = \frac{1}{2\sqrt{2}}\log\frac{\sqrt{1+x^2}+\sqrt{2}x}{\sqrt{1+x^2}-\sqrt{2}x}$$

$$= \frac{1}{2\sqrt{2}}\log\frac{(\sqrt{1+x^2}+\sqrt{2}x)^2}{1+x^2-2x^2} = \frac{1}{\sqrt{2}}\log\frac{\sqrt{1+x^2}+\sqrt{2}x}{\sqrt{1-x^2}}.$$

(35.) $\displaystyle\int\frac{dx}{(1+x)\sqrt{1-x-x^2}}.$ Let $1+x = \frac{1}{z}$

$$x = \frac{1}{z}-1 \qquad dx = -\frac{dz}{z^2}$$

$$1-x-x^2 = 1-\frac{1}{z}+1-\frac{1}{z^2}+\frac{2}{z}-1 = 1+\frac{1}{z}-\frac{1}{z^2},$$

D 2

$$\therefore \int \frac{dx}{(1+x)\sqrt{1-x-x^2}} = -\int \frac{dz}{z^2 \frac{1}{z}\sqrt{1+\frac{1}{z}-\frac{1}{z^2}}}$$

$$= -\int \frac{dz}{\sqrt{z^2+z-1}} = -\int \frac{dz}{\sqrt{\left(z+\frac{1}{2}\right)^2-\frac{5}{4}}}$$

$$= -\log\left\{\sqrt{z^2+z-1}+z+\frac{1}{2}\right\}$$

$$= -\log\left\{\frac{\sqrt{1-x-x^2}}{1+x}+\frac{1}{1+x}+\frac{1}{2}\right\}$$

$$= -\log\left\{\frac{2\sqrt{1-x-x^2}+x+3}{2(1+x)}\right\}$$

$$= \log\left\{\frac{2(1+x)}{2\sqrt{1-x-x^2}+x+3}\right\}$$

$$= \log\left\{\frac{2(1+x)(x+3)-2\sqrt{1-x-x^2}}{x^2+6x+9-4+4x+4x^2}\right\}.$$

$$= \log\frac{2(1+x)(x+3-2\sqrt{1-x-x^2})}{5(1+x)^2}$$

$$= \log\frac{x+3-2\sqrt{1-x-x^2}}{1+x}+c.$$

(36.) $$\int \frac{dx}{\sqrt{1+2x-x^2}} = \int \frac{dx}{\sqrt{2-(1-x)^2}}$$

$$= -\int \frac{d(1-x)}{\sqrt{2-(1-x)^2}} = \cos^{-1}\frac{1-x}{\sqrt{2}}.$$

(37.) $\displaystyle\int \frac{d\,x}{(a + b\,x^2)^{\frac{3}{2}}}.$ Let $x = \dfrac{1}{z}, \quad d\,x = -\dfrac{d\,z}{z^2}$

$$= -\int \frac{d\,z}{z^2 \left(a + \dfrac{b}{z^2} \right)^{\frac{3}{2}}} = -\int \frac{z\,d\,z}{(a\,z^2 + b)^{\frac{3}{2}}}$$

$$= \frac{1}{a} \frac{1}{(a\,z^2 + b)^{\frac{1}{2}}} = \frac{x}{a\,(a + b\,x^2)^{\frac{1}{2}}}.$$

(38.) $\displaystyle\int x\,d\,x \sqrt{\frac{1+x}{1-x}} = \int \frac{x\,dx}{\sqrt{1-x^2}} + \int \frac{x^2\,dx}{\sqrt{1-x^2}}$

$$\int x\, \frac{x\,dx}{\sqrt{1-x^2}} = -x\sqrt{1-x^2} + \int \sqrt{1-x^2}\,dx$$

$$= -x\sqrt{1-x^2} + \int \frac{dx}{\sqrt{1-x^2}}$$

$$-\int \frac{x^2\,dx}{\sqrt{1-x^2}}$$

$$\int x^2\, \frac{dx}{\sqrt{1-x^2}} = -\frac{x\sqrt{1-x^2}}{2} + \frac{1}{2}\,\sin^{-1} x,$$

$$\therefore \int x\,d\,x \sqrt{\frac{1+x}{1-x}} = -\sqrt{1-x^2}$$

$$-\frac{x}{2}\sqrt{1-x^2} + \frac{1}{2}\,\sin^{-1} x$$

$$= \frac{1}{2}\,\sin^{-1} x - (2 + x)\,\frac{\sqrt{1-x^2}}{2}.$$

(39.) $\qquad \displaystyle\int \frac{\sqrt{x}\, dx}{\sqrt{a^3 - x^3}}.$ Let $x^3 = z^2$

$$x = z^{\frac{2}{3}}, \; dx = \frac{2}{3}\, z^{-\frac{1}{3}}\, dz, \; \sqrt{x} = z^{\frac{1}{3}},$$

$$\therefore \int \frac{\sqrt{x}\, dx}{\sqrt{a^3 - x^3}} = \frac{2}{3} \int \frac{dz}{\sqrt{a^3 - z^2}} = \frac{2}{3} \sin^{-1} \frac{z}{a^{\frac{3}{2}}}$$

$$= \frac{2}{3} \sin^{-1} \frac{x^{\frac{3}{2}}}{a^{\frac{3}{2}}}$$

Let $\theta = \sin^{-1} \dfrac{x^{\frac{3}{2}}}{a^{\frac{3}{2}}}, \quad \sin \theta = \dfrac{x^{\frac{3}{2}}}{a^{\frac{3}{2}}},$

$$\tan \theta = \frac{\sin \theta}{\sqrt{1 - \sin^2 \theta}} = \frac{x^{\frac{3}{2}}}{\sqrt{a^3 - x^3}},$$

$$\therefore \int \frac{\sqrt{x}\, dx}{\sqrt{a^3 - x^3}} = \frac{2}{3} \tan^{-1} \sqrt{\frac{x^3}{a^3 - x^3}}.$$

(40.) $\qquad \displaystyle\int \frac{dx}{(2ax + x^2)^{\frac{3}{2}}} = \int \frac{dx}{\left((x+a)^2 - a^2\right)^{\frac{3}{2}}}.$

Let $x + a = \dfrac{1}{z}, \; dx = -\dfrac{dz}{z^2}$

$$= -\int \frac{dz}{z^2 \left(\dfrac{1}{z^2} - a^2\right)^{\frac{3}{2}}} = -\int \frac{z\, dz}{(1 - a^2 z^2)^{\frac{3}{2}}}.$$

$$= -\frac{1}{a^3}\frac{1}{\sqrt{1-a^2 z^2}} = -\frac{1}{a^3}\frac{1}{\sqrt{1-\dfrac{a^2}{(x+a)^2}}} \quad ,$$

$$= -\frac{x+a}{a^2\sqrt{2\,a\,x + x^2}}.$$

(41.) $\displaystyle\int\frac{d\,x}{x^3\sqrt{2\,a\,x - x^2}}.$ Let $x = \dfrac{1}{z}$ $d\,x = -\dfrac{d\,z}{z^2}$

$$= -\int\frac{d\,z}{z^2\dfrac{1}{z^2}\sqrt{\dfrac{2\,a}{z}-\dfrac{1}{z^2}}} = -\int\frac{z\,d\,z}{\sqrt{2\,a\,z - 1}}.$$

$$= -\frac{1}{a}\,z\sqrt{2\,a\,z-1} + \int\sqrt{2\,a\,z-1}\,.\,d\,z$$

$$= -\frac{1}{a}\,z\sqrt{2\,a\,z-1} + \int\frac{2\,z\,d\,z}{\sqrt{2\,a\,z-1}} - \frac{1}{a}\int\frac{d\,z}{\sqrt{2\,a\,z-1}},$$

$$\therefore \int\frac{d\,x}{x^3\sqrt{2\,a\,x - x^2}} = -\frac{1}{3\,a}\,z\sqrt{2\,a\,z-1} - \frac{1}{3\,a^2}\sqrt{2\,a\,z-1}.$$

$$= \frac{1}{3\,a}\left(\frac{1}{x^2}-\frac{1}{a\,x}\right)\sqrt{2\,a\,x - x^2}.$$

(42.) $\displaystyle\int x^7\sqrt{\frac{1+x^2}{1-x^2}}\,d\,x = \int\frac{x^7\,d\,x}{\sqrt{1-x^4}} + \int\frac{x^5\,d\,x}{\sqrt{1-x^4}},$

$$\int\frac{x^5\,d\,x}{\sqrt{1-x^4}} = \int x^3\frac{x^3\,d\,x}{\sqrt{1-x^4}}$$

$$= -\frac{x^2}{2}\sqrt{1-x^4} + \int x\sqrt{1-x^4}\,d\,x$$

$$= -\frac{x^2}{2} + \int\frac{x\,d\,x}{\sqrt{1-x^4}} - \int\frac{x^5\,d\,x}{\sqrt{1-x^4}},$$

$$\therefore \int \frac{x^5\,dx}{\sqrt{1-x^4}} = -\frac{x^2\sqrt{1-x^4}}{4} - \frac{1}{4}\int \frac{d\,(-x^2)}{\sqrt{1-(-x^2)^2}},$$

$$= -\frac{x^3}{4}\sqrt{1-x^4} + \frac{1}{4}\cos^{-1}(-x^2),$$

$$\therefore \int x^3 \sqrt{\frac{1+x^2}{1-x^2}}\,dx$$

$$= -\left(\frac{1}{2}+\frac{x^2}{4}\right)\sqrt{1-x^4} + \frac{1}{4}\cos^{-1}(-x^2).$$

Let $\theta = \dfrac{1}{4}\cos^{-1}(-x^2)$

$$\cos 4\,\theta = -x^2$$

$$\tan 2\,\theta = \sqrt{\frac{1-\cos 4\,\theta}{1+\cos 4\,\theta}} = \sqrt{\frac{1+x^2}{1-x^2}}$$

$$\theta = \frac{1}{2}\tan^{-1}\sqrt{\frac{1+x^2}{1-x^2}}$$

$$\therefore \int x^3 \sqrt{\frac{1+x^2}{1-x^2}}\,dx$$

$$= \frac{1}{2}\tan^{-1}\sqrt{\frac{1+x^2}{1-x^2}} - \frac{1}{4}(2+x^2)\sqrt{1-x^4}.$$

(43.) $\displaystyle\int \frac{d\,x}{x}\sqrt{\frac{a^2-c^2\,x^2}{a^2-x^2}}.$ Let $y^2 = \dfrac{a^2-c^2\,x^2}{a^2-x^2}$

$$a^2\,y^2 - x^2\,y^2 = a^2 - c^2\,x^2$$

$$x^2 = \frac{a^2\,(1-y^2)}{c^2-y^2},$$

$$2\,x\,dx = a^2\left\{\frac{-2y\,(c^2-y^2)+2\,y\,(1-y^2)}{(c^2-y^2)^2}\right\}\,dy$$

$$x \, dx = \frac{a^2 \, y \, (1 - c^2) \, dy}{(c^2 - y^2)^2}$$

$$\frac{dx}{x} = \frac{a^2 \, y \, (1 - c^2) \, dy}{(c^2 - y^2)^2} \cdot \frac{c^2 - y^2}{a^2 \, (1 - y^2)} = \frac{y \, (1 - c^2) \, dy}{(1 - y^2) \, (c^2 - y^2)},$$

$$\therefore \int \frac{dx}{x} \sqrt{\frac{a^2 - c^2 \, x^2}{a^2 - x^2}} = \int \frac{y^2 \, (1 - c^2) \, dy}{(1 - y^2) \, (c^2 - y^2)}$$

$$= -\frac{1}{2} \int \frac{1}{1 + y} - \frac{1}{2} \int \frac{1}{1 - y} + \frac{c}{2} \int \frac{1}{c + y} + \frac{c}{2} \int \frac{1}{c - y}$$

$$= \frac{1}{2} \log \frac{1 - y}{1 + y} + \frac{c}{2} \log \frac{c + y}{c - y}$$

$$= \log \sqrt{\left(\frac{1 - y}{1 + y}\right) \left(\frac{c + y}{c - y}\right)^c}.$$

(44.)
$$\int \frac{x^2 + 1}{x^4 - 1} \frac{dx}{\sqrt{1 - a \, x^2 + x^4}}$$

$$= \int \frac{1 + x^{-2}}{x - x^{-1}} \frac{dx}{\sqrt{x^2 - a + x^{-2}}}$$

$$= \int \frac{d \, (x - x^{-1})}{(x - x^{-1}) \sqrt{(x^{-1} - x)^2 - (a - 2)}}$$

$$= \frac{1}{\sqrt{a - 2}} \sec^{-1} \frac{x^{-1} - x}{\sqrt{a - 2}} = \frac{1}{\sqrt{a - 2}} \sec^{-1} \frac{x^2 - 1}{x \sqrt{a - 2}}$$

$$= \frac{1}{\sqrt{a - 2}} \cos^{-1} \frac{x \sqrt{a - 2}}{x^2 - 1}.$$

FORMULÆ OF REDUCTION.

$$\int \frac{x^n \, dx}{\sqrt{2\,ax - x^2}},$$

$$\int \frac{x^n \, dx}{\sqrt{2\,ax - x^2}} = -\int \frac{x^{n-1}\,(a - x)\,dx}{\sqrt{2\,ax - x^2}} + a\int \frac{x^{n-1}\,dx}{\sqrt{2\,ax - x^2}},$$

Making $p = x^{n-1}$, $dq = \dfrac{(a - x)\,dx}{\sqrt{2\,ax - x^2}}$,

and $\therefore dp = (n-1)\,x^{n-2}\,dx$, $q = \sqrt{2\,ax - x^2}$,

we have $\int p\,dq = pq - \int q\,dp$, $\therefore \int \dfrac{x^{n-1}\,(a - x)}{\sqrt{2\,ax - x^2}}$

$$= x^{n-1}\sqrt{2\,ax - x^2} - (n-1)\int x^{n-2}\sqrt{2\,ax - x^2}\,dx$$

$$= x^{n-1}\sqrt{2\,ax - x^2} - (n-1) \cdot 2\,a\int \frac{x^{n-1}\,dx}{\sqrt{2\,ax - x^2}}$$

$$+ (n-1)\int \frac{x^n \, dx}{\sqrt{2\,ax - x^2}}.$$

Substituting this value in the original expression, we have

$$\int \frac{x^n \, dx}{\sqrt{2\,ax - x^2}} = -x^{n-1}\sqrt{2\,ax - x^2}$$

$$+ 2\,a \cdot (n-1)\int \frac{x^{n-1}\,dx}{\sqrt{2\,ax - x^2}} - (n-1)\int \frac{x^n \, dx}{\sqrt{2\,ax - x^2}}$$

$$+ a\int \frac{x^{n-1}\,dx}{\sqrt{2\,ax - x^2}},$$

$$\therefore \left(1 + n - 1\right) \int \frac{x^n \, dx}{\sqrt{2\,ax - x^2}} = -\, x^{n-1} \sqrt{2\,ax - x^2}$$

$$+\, a\,(2n - 1) \int \frac{x^{n-1} \, dx}{\sqrt{2\,ax - x^2}},$$

$$\therefore \int \frac{x^n \, dx}{\sqrt{2\,ax - x^2}} = \frac{-x^{n-1}\sqrt{2\,ax - x^2}}{n}$$

$$+\, \frac{a\,(2n - 1)}{n} \int \frac{x^{n-1} \, dx}{\sqrt{2\,ax - x^2}}.$$

$$\int \frac{x^n \, dx}{\sqrt{2\,ax + x^2}},$$

$$\int \frac{x^n \, dx}{\sqrt{2\,ax + x^2}} = \int x^{n-1} \frac{(a + x)\, dx}{\sqrt{2\,ax + x^2}} - a \int \frac{x^{n-1} \, dx}{\sqrt{2\,ax + x^2}}$$

$$\int \frac{x^{n-1}\,(a + x)}{\sqrt{2\,ax + x^2}}. \qquad \text{This becomes, if } p = x^{n-1}$$

$$dq = \frac{(a + x)\, dx}{\sqrt{2\,ax + x^2}}, \quad \therefore q = \sqrt{2\,ax + x^2}, \int p\,dq = pq - \int q\,dp,$$

$$\therefore \int \frac{x^{n-1}\,(a + x)}{\sqrt{2\,ax + x^2}} = x^{n-1}\sqrt{2\,ax + x^2}$$

$$-\,(n - 1) \int x^{n-2}\sqrt{2\,ax + x^2} \, dx = x^{n-1}\sqrt{2\,ax + x^2}$$

$$-\, 2 . a\,(n - 1) \int \frac{x^{n-1} \, dx}{\sqrt{2\,ax + x^2}} - (n - 1) \int \frac{x^n \, dx}{\sqrt{2\,ax + x^2}}.$$

Substituting this in the original expression,

$$\int \frac{x^n\,dx}{\sqrt{2\,ax+x^2}} = x^{n-1}\sqrt{2\,ax+x^2} - 2\,a\,(n-1)\int \frac{x^{n-1}\,dx}{\sqrt{2\,ax+x^2}}$$

$$- (n-1)\int \frac{x^n\,dx}{\sqrt{2\,ax+x^2}} - a\int \frac{x^{n-1}\,dx}{\sqrt{2\,ax+x^2}}$$

$$= \frac{x^{n-1}\sqrt{2\,ax+x^2}}{n} - \frac{a(2n-1)}{n}\int \frac{x^{n-1}\,dx}{\sqrt{2\,ax+x^2}}.$$

$$\int \frac{dx}{x^n\sqrt{2\,ax-x^2}}$$

$$\int \frac{dx}{x^{n-1}\sqrt{2\,ax-x^2}} = -\int \frac{dx\,(a-x)}{x^n\sqrt{2\,ax-x^2}}$$

$$+ a\int \frac{dx}{x^n\sqrt{2\,ax-x^2}} \;\ldots\ldots\ldots\; (a.)$$

Now, making $p = \dfrac{1}{x^n}$ and $dq = \dfrac{(a-x)\,dx}{\sqrt{2\,ax-x^2}}$,

and $\therefore\; dp = -n\,a^{-n-1}\,dx,\quad q = \sqrt{2\,ax-x^2},$

there results

$$\int \frac{(a-x)\,dx}{x^n\sqrt{2\,ax-x^2}} = \frac{\sqrt{2\,ax-x^2}}{x^n} + n\int \frac{\sqrt{2\,ax-x^2}\,.\,dx}{x^{n+1}}$$

$$= \frac{\sqrt{2\,ax-x^2}}{x^n} + 2\,an\int \frac{dx}{x^n\sqrt{2\,ax-x^2}}$$

$$- n\int \frac{dx}{x^{n-1}\sqrt{2\,ax-x^2}}.$$

Substituting this in the equation (a) we obtain

$$\int\frac{dx}{x^{n-1}\sqrt{2ax-x^2}} = -\frac{\sqrt{2ax-x^2}}{x^n} - 2an\int\frac{dx}{x^n\sqrt{2ax-x^2}}$$

$$+ n\int\frac{dx}{x^{n-1}\sqrt{2ax-x^2}} + a\int\frac{dx}{x^n\sqrt{2ax-x^2}},$$

$$\therefore\ a(2n-1)\int\frac{dx}{x^n\sqrt{2ax-x^2}} = -\frac{\sqrt{2ax-x^2}}{x^n}$$

$$+ (n-1)\int\frac{dx}{x^{n-1}\sqrt{2ax-x^2}},$$

$$\therefore\ \int\frac{dx}{x^n\sqrt{2ax-x^2}} = \frac{\sqrt{2ax-x^2}}{a(2n-1)x^n}$$

$$+ \frac{n-1}{a(2n-1)}\int\frac{dx}{x^{n-1}\sqrt{2ax-x^2}}.$$

$$\int\frac{dx}{x^n\sqrt{2ax+x^2}}.$$

$$\int\frac{dx}{x^{n-1}\sqrt{2ax+x^2}} = \int\frac{(a+x)dx}{x^n\sqrt{2ax+x^2}} - a\int\frac{dx}{x^n\sqrt{2ax+x^2}}.$$

If $p = x^{-n}$ and $dq = \frac{(a+x)dx}{\sqrt{2ax+x^2}}$,

We have $\int\frac{(a+x)dx}{x^n\sqrt{2ax+x^2}} = \frac{\sqrt{2ax+x^2}}{x^n} + n\int\frac{\sqrt{2ax+x^2}\,dx}{x^{n+1}}$

$$= \frac{\sqrt{2ax+x^2}}{x^n} + 2an\int\frac{dx}{x^n\sqrt{2ax+x^2}},$$

$$+ n\int\frac{dx}{x^{n-1}\sqrt{2ax+x^2}},$$

and by substitution in the first expression there results :

$$\int\frac{dx}{x^{n-1}\sqrt{2ax+x^2}} = \frac{\sqrt{2ax+x^2}}{x^n},$$

$$+ (2an - a) \int \frac{dx}{x^n \sqrt{2ax + x^2}} + n \int \frac{dx}{x^{n-1} \sqrt{2ax + x^2}},$$

$$\therefore \frac{(1 - n)}{a(2n - 1)} \int \frac{dx}{x^{n-1}\sqrt{2ax + x^2}} - \frac{\sqrt{2ax + x^2}}{x^n a(2n - 1)}$$

$$= \int \frac{dx}{x^n \sqrt{2ax + x^2}}.$$

$$\int \frac{dx}{(a^2 + x^2)^n}. \quad \text{We have} \quad \frac{1}{(a^2 + x^2)^n}$$

$$= \frac{1}{a^2} \cdot \frac{a^2}{(a^2 + x^2)^n} = \frac{1}{a^2} \left\{ \frac{a^2 + x^2}{(a^2 + x^2)^n} - \frac{x^2}{(a^2 + x^2)^n} \right\}$$

$$= \frac{1}{a^2} \cdot \frac{1}{(a^2 + x^2)^{n-1}} - \frac{1}{a^2} \frac{x^2}{(a^2 + x^2)^n}.$$

$$\therefore \int \frac{dx}{(a^2 + x^2)^n} = \frac{1}{a^2} \int \frac{dx}{(a^2 + x^2)^{n-1}} - \frac{1}{a^2} \int \frac{x^2 dx}{(a^2 + x^2)^n}.$$

But $\int \frac{x^2 dx}{(a^2 + x^2)^n} = \int x \frac{x}{(a^2 + x^2)^n} dx = \int p \, dq,$

if $p = x$ and $dq = \frac{x \, dx}{(a^2 + x^2)^n};$

$$\therefore \int \frac{x^2 dx}{(a^2 + x^2)^n} = \frac{- x}{(2n - 2)(a^2 + x^2)^{n-1}}$$

$$+ \frac{1}{2n - 2} \int \frac{dx}{(a^2 + x^2)^{n-1}};$$

$$\therefore \int \frac{d\,x}{(a^2 + x^2)^n} = \frac{1}{a^2}\frac{x}{(2\,n - 2)\,(a^2 + x^2)^{n-1}}$$

$$- \frac{1}{a^2 \cdot (2n - 2)} \int \frac{d\,x}{(a^2 + x^2)^{n-1}} + \frac{1}{a^2} \int \frac{d\,x}{(a^2 + x^2)^{n-1}}$$

$$= \frac{1}{a^2}\frac{x}{(2\,n - 2)\,(a^2 + x^2)^{n-1}} + \frac{2n - 3}{a^2\,(2\,n - 2)} \int \frac{d\,x}{(a^2 + x^2)^{n-1}}.$$

Also for $\int \dfrac{x^m\,d\,x}{(a^2 + x^2)^n} = \int x^{m-1}\,d\,x \dfrac{x}{(a^2 + x^4)^n}.$ We have,

if $p = x^{m-1}$, $d\,q = x\,(a^2 + x^2)^{-n}\,d\,x$.

$$\int \frac{x^m\,d\,x}{(a^2 + x^4)^n}$$

$$= \frac{1}{2 - 2n}\frac{x^{m-1}}{(a^2 + x^2)^{n-1}} + \frac{m - 1}{m - 2} \int \frac{x^{m-2}\,d\,x}{(a^2 + x^2)^{n-1}}$$

$$d\,u = (a^2 - x^2)^{\frac{n}{2}}\,d\,x$$

$$(a^2 - x^2)^{\frac{n}{2}} = a^2\,(a^2 - x^2)^{\frac{n-2}{2}} - x^2\,(a^2 - x^2)^{\frac{n-2}{2}};$$

$$\therefore\ u = a^2 \int (a^2 - x^2)^{\frac{n-2}{2}}\,d\,x - \int x \cdot x\,(a^2 - x^2)^{\frac{n-2}{2}}\,d\,x$$

$$= a^2 \int (a^2 - x^2)^{\frac{n-2}{2}}\,d\,x + \frac{x\,(a^2 - x^2)^{\frac{n}{2}}}{n} - \frac{u}{n};$$

$$\therefore\ u = \frac{x\,(a^2 - x^2)^{\frac{n}{2}}}{n + 1} + \frac{n\,a^2}{n + 1} \int (a^2 - x^2)^{\frac{n-2}{2}}\,d\,x.$$

$$\int \frac{x^n\,dx}{\sqrt{2\,a\,x-x^2}} = -\frac{x^{n-1}\sqrt{2\,a\,x-x^2}}{n}$$

$$+ a\left(\frac{2n-1}{n}\right)\int \frac{x^{n-1}\,dx}{\sqrt{2\,a\,x-x^2}}.$$

If we make $n=2$ this expression becomes

$$\int \frac{x^2\,dx}{\sqrt{2\,a\,x-x^2}} = -\frac{x\sqrt{2\,a\,x-x^2}}{2} + \frac{3\,a}{2}\int \frac{x\,dx}{\sqrt{2\,a\,x-x^2}}.$$

Also $\displaystyle\int \frac{x\,dx}{\sqrt{2\,a\,x-x^2}} = -\sqrt{2\,a\,x-x^2} + a\ \mathrm{ver}\ \sin^{-1}\left(\frac{x}{a}\right);$

$$\therefore\quad \frac{x^2\,dx}{\sqrt{2\,a\,x-x^2}} = -\frac{x\sqrt{2\,a\,x-x^2}}{2} - \frac{a}{2}\sqrt{2\,a\,x-x^2}$$

$$+ \frac{3\,a^2}{2}\ \mathrm{ver}\ \sin^{-1}\left(\frac{x}{a}\right)$$

$$= -\frac{(x+3a)}{2}\sqrt{2\,a\,x-x^2} + \frac{3\,a^2}{2}\ \mathrm{ver}\ \sin^{-1}\left(\frac{x}{a}\right)$$

$$\int \frac{dx}{(a^2+x^2)^n} = \frac{1}{2n-2}\cdot\frac{x}{a^2\,(a^2+x^2)^{n-1}} +$$

$$\frac{2n-3}{2n-2}\cdot\frac{1}{a^2}\cdot\int \frac{dx}{(a^2+x^2)^{n-1}}.$$

If $n=4$ we have $\displaystyle\int \frac{dx}{(a^2+x^2)^4} =$

$$\frac{1}{6\,a^2}\cdot\frac{x}{(a^2+x^2)^3} + \frac{5}{6\,a^2}\int \frac{dx}{(a^2+x^2)^3},$$

$$\int \frac{dx}{(a^2 + x^2)^3} = \frac{1}{4a^2} \cdot \frac{x}{(a^2 + x^2)^2} + \frac{3}{4a^2} \int \frac{dx}{(a^2 + x^2)^2}.$$

Also, $\int \frac{dx}{(a^2 + x^2)^2} = \frac{1}{2} \frac{x}{a^2 (a^2 + x^2)} + \frac{1}{2} \tan^{-1} \frac{x}{a}$,

∴ the required integral is

$$\frac{1}{6a^2} \cdot \frac{x}{(a^2 + x^2)^3} + \frac{5}{6 \cdot 4a^4} \frac{x}{(a^2 + x^2)^2}$$

$$+ \frac{5 \cdot 3}{6 \cdot 4 \cdot 2a^6} \cdot \frac{x}{a^2 + x^2} + \frac{5 \cdot 3}{6 \cdot 4 \cdot 2} \frac{1}{a^7} \cdot \tan^{-1} \left(\frac{x}{a} \right).$$

$$\int \frac{dx}{x^n (1 + x^2)^{\frac{1}{2}}}.$$

The formula of reduction is

$$- \frac{1}{(n-1)} \frac{(x^2 + 1)^{\frac{1}{2}}}{x^{n-1}} - \frac{n-2}{n-1} \int \frac{dx}{x^{n-2} (1 + x^2)^{\frac{1}{2}}}.$$

If $n = 6$, we have $\int \frac{dx}{x^6 (1 + x^2)^{\frac{1}{2}}}$

$$= - \frac{1}{5} \cdot \frac{(x^2 + 1)^{\frac{1}{2}}}{x^5} - \frac{4}{5} \cdot \int \frac{dx}{x^4 (1 + x^2)^{\frac{1}{2}}}$$

$$\int \frac{dx}{x^4 (1 + x^2)^{\frac{1}{2}}} = - \frac{1}{3} \cdot \frac{(x^2 + 1)^{\frac{1}{2}}}{x^3} - \frac{2}{3} \int \frac{dx}{x^2 (x^2 + 1)^{\frac{1}{2}}},$$

$$\int \frac{dx}{x^2 (x^2 + 1)^{\frac{1}{2}}} = \frac{- (1 + x^2)^{\frac{1}{2}}}{x}.$$

And by reduction this becomes

$$\int \frac{dx}{x^6 (x^2 + 1)^{\frac{1}{2}}} = -\frac{1}{5} \frac{(1 + x^2)^{\frac{1}{2}}}{x^5} + \frac{4}{3 \cdot 5} \frac{(1 + x^2)^{\frac{1}{2}}}{x^3}$$

$$-\frac{4 \cdot 2}{5 \cdot 3} \cdot \frac{(1 + x^2)^{\frac{1}{2}}}{x}.$$

$$\int \frac{dx}{x^n \sqrt{(x^2 - 1)}}.$$

The formula of reduction is

$$\int \frac{dx}{x^n (x^2 - 1)^{\frac{1}{2}}} = \frac{1}{n - 1} \cdot \frac{(x^2 - 1)^{\frac{1}{2}}}{x^{n-1}} + \frac{n - 2}{n - 1} \int \frac{dx}{x^{n-2} (x^2 - 1)^{\frac{1}{2}}}.$$

If $n = 6$, this becomes

$$\int \frac{dx}{x^6 (x^2 - 1)^{\frac{1}{2}}} = \frac{1}{5} \cdot \frac{(x^2 - 1)^{\frac{1}{2}}}{x^5} + \frac{4}{5} \int \frac{dx}{x^4 (x^2 - 1)^{\frac{1}{2}}},$$

also $\int \frac{dx}{x^4 (x^2 - 1)^{\frac{1}{2}}} = \frac{1}{3} \cdot \frac{(x^2 - 1)^{\frac{1}{2}}}{x^3} + \frac{2}{3} \int \frac{dx}{x^2 (x^2 - 1)^{\frac{1}{2}}}$

$$\int \frac{dx}{x^2 (x^2 - 1)^{\frac{1}{2}}} = \frac{(x^2 - 1)^{\frac{1}{2}}}{x}.$$

Therefore, we have the required integral

$$= \frac{1}{5} \cdot \frac{(x^2 - 1)^{\frac{1}{2}}}{x^5} + \frac{4}{3 \cdot 5} \frac{(x^2 - 1)^{\frac{1}{2}}}{x^3} + \frac{2 \cdot 4}{3 \cdot 5} \frac{(x^2 - 1)^{\frac{1}{2}}}{x}.$$

CHAPTER IV.

(1.) $\dfrac{d\,u}{d\,x} = x^2 (\log x)^2$ $\quad (\log x)^2 = p, \quad \dfrac{d\,q}{d\,x} = x^2$

$$\dfrac{d\,p}{d\,x} = 2\log x \cdot \dfrac{1}{x}, \quad q = \dfrac{x^3}{3}$$

$$u = \dfrac{x^3 (\log x)^2}{3} - \dfrac{2}{3} \int x^2\, dx \log x$$

$$\int x^2\, dx \log x \begin{cases} p = \log x & \dfrac{d\,q}{d\,x} = x^2 \\[2mm] \dfrac{d\,p}{d\,x} = \dfrac{1}{x} & q = \dfrac{x^3}{3} \end{cases}$$

$$= \dfrac{x^3 \log x}{3} - \dfrac{1}{3}\int x^2\, dx = \dfrac{x^3 \log x}{3} - \dfrac{x^3}{9},$$

$$\therefore u = \dfrac{x^3 (\log x)^2}{3} - \dfrac{2x^3 \log x}{9} + \dfrac{2x^3}{27}$$

$$= \dfrac{x^3}{3}\left\{ (\log x)^2 - \dfrac{2}{3}\log x + \dfrac{2}{9}\right\}.$$

(2.) $\displaystyle\int \dfrac{x^3\, dx}{\sqrt{\log x}}$ $\begin{cases} p = (\log x)^{-\frac{1}{2}} & dq = x^3\, dx \\[2mm] dp = -\dfrac{1}{2}(\log x)^{-\frac{3}{2}} \cdot \dfrac{d\,x}{x} & q = \dfrac{x^4}{4} \end{cases}$

$$= \dfrac{x^4}{4}(\log x)^{-\frac{1}{2}} + \dfrac{1}{2\cdot 4}\int x^3\, dx\, (\log x)^{-\frac{3}{2}}$$

Similarly integrating by parts,

$$\int x^3 \, dx \, (\log x)^{-\frac{3}{2}} = \frac{x^4}{4} (\log x)^{-\frac{3}{2}} + \frac{3}{2 \cdot 4} \int x^3 \, dx \, (\log x)^{-\frac{5}{2}},$$

$$\int x^3 \, dx \, (\log x)^{-\frac{5}{2}} = \frac{x^4}{4} (\log x)^{-\frac{5}{2}} + \frac{5}{2 \cdot 4} \int x^3 \, dx \, (\log x)^{-\frac{7}{2}},$$

$$\int x^3 \, dx \, (\log x)^{-\frac{7}{2}} = \frac{x^4}{4} (\log x)^{-\frac{7}{2}} + \frac{7}{2 \cdot 4} \int x^3 \, dx \, (\log x)^{-\frac{9}{2}},$$

$$\therefore u = \frac{x^4}{4 \sqrt{\log x}}$$

$$+ \frac{1}{8} \left\{ \frac{x^4}{4 (\log x)^3} + \frac{3}{8} \left(\frac{x^4}{4 (\log x)^{\frac{5}{2}}} + \frac{5}{8} \cdot \frac{x^4}{4 (\log x)} \right) \right\},$$

$$u =$$

$$\frac{x^4}{4 \sqrt{\log x}} \left\{ 1 + \frac{1}{8 \log x} + \frac{3}{(8 \log x)^2} + \frac{3 \cdot 5}{(8 \log x)^3} + \&c. \right\}.$$

$$(3.) \int \frac{x^4 \, dx}{(\log x)^3} = \int x^5 \, \frac{\frac{dx}{x}}{(\log x)^3},$$

$$p = x^5 \qquad dq = \frac{\frac{dx}{x}}{(\log x)^3}$$

$$dp = 5 x^4 \, dx \qquad q = - \frac{1}{2 (\log x)^2},$$

$$\therefore u = - \frac{x^5}{2 (\log x)^2} + \frac{5}{2} \int \frac{x^4 \, dx}{(\log x)^2}.$$

In like manner,

$$\int \frac{x^4\,dx}{2\,(\log x)^2} = -\frac{x^5}{\log x} + 5\int \frac{x^4\,dx}{\log x},$$

$$\therefore u = -\frac{x^5}{2\,(\log x)^2} - \frac{5\,x^5}{2\,\log x} + \frac{25}{2}\int \frac{x^4\,dx}{\log x}.$$

$$(4.)\ \int a^x x^3\,dx \begin{cases} p = x^3 & \dfrac{dq}{dx} = a^x \\[2ex] dp = 3x^2\,dx & q = \dfrac{a^x}{A} \end{cases}$$

$$= \frac{x^3 a^x}{A} - \frac{3}{A}\int a^x x^2\,dx$$

$$\int a^x x^2\,dx = \frac{a^x x^2}{A} - \frac{2}{A}\int a^x x\,dx$$

$$\int a^x x\,dx = \frac{a^x x}{A} - \frac{1}{A}\int a^x\,dx = \frac{a^x x}{A} - \frac{a^x}{A^2},$$

$$\therefore \int a^x x^3\,dx = \frac{a^x x^3}{A} - \frac{3}{A}\left\{\frac{a^x x^2}{A} - \frac{2}{A}\left(\frac{a^x x}{A} - \frac{a^x}{A}\right)\right\}$$

$$= a^x\left\{\frac{x^3}{A} - \frac{3x^2}{A^2} + \frac{6x}{A^3} - \frac{6}{A^4}\right\}.$$

$$(5.)\ \int x^3\,(\log x)^3\,dx \quad p = (\log x)^3 \quad dq = x^3 dx$$

$$dp = 3\,(\log x)^2\,\frac{dx}{x} \qquad q = \frac{x^4}{4}$$

$$\int x^3\,(\log x)^3\,dx = \frac{x^4}{4}\,(\log x)^3 - \frac{3}{4}\int x^3\,(\log x)^2\,dx$$

similarly, $\displaystyle\int x^3\,(\log x)^2\,dx = \frac{x^4}{4}\,(\log x)^2 - \frac{2}{4}\int x^3\,\log x\,dx$

$$\int x^3 \, dx \log x = \frac{x^4}{4} \log x - \frac{1}{4} \int x^3 \, dx$$

$$= \frac{x^4}{4} \log x - \frac{1}{16} x^4,$$

$$\therefore \int x^3 (\log x)^3 \, dx =$$

$$\frac{x^4}{4} \left\{ (\log x)^3 - \frac{3}{4} (\log x)^2 + \frac{3.2}{4^2} \log x - \frac{3.2}{4^3} \right\}.$$

(6.) $\int e^x x^4 \, dx = e^x x^4 - 4 \int e^x x^3 \, dx$

$$\int e^x x^3 \, dx = e^x x^3 - 3 \int e^x x^2 \, dx$$

$$\int e^x x^2 \, dx = e^x x^2 - 2 \int e^x x \, dx$$

$$\int e^x x \, dx = e^x x - \int e^x \, dx = e^x x - e^x,$$

$$\therefore \int e^x x^4 \, dx = e^x (x^4 - 4x^3 + 12x^2 - 24x + 24).$$

(7.) $\int e^{-x} x^3 \, dx = - e^{-x} x^3 + 3 \int e^{-x} x^2 \, dx$

$$\int e^{-x} x^2 \, dx = - e^{-x} x^2 + 2 \int e^{-x} x \, dx$$

$$\int e^{-x} x \, dx = - e^{-x} x + \int e^{-x} \, dx = - e^{-x} x - e^{-x},$$

$$\therefore \int e^{-x} x^3 \, dx = - e^{-x} (x^3 + 3x^2 + 6x + 6).$$

$$(8.) \int \frac{x\,dx}{\sqrt{1+x^2}} \log x \quad \left\{ \begin{array}{ll} p = \log x & \dfrac{dq}{dx} = \dfrac{x}{\sqrt{1+x^2}} \\[3mm] \dfrac{dp}{dx} = \dfrac{1}{x} & q = \sqrt{1+x^2} \end{array} \right\}$$

$$= \sqrt{1+x^2}\,\log x - \int \frac{\sqrt{1+x^2}\,dx}{x}$$

$$= \sqrt{1+x^2}\,\log x - \int \frac{dx}{x\sqrt{1+x^2}} - \int \frac{x\,dx}{\sqrt{1+x^2}}$$

$$= \sqrt{1+x^2}\,\log x - \log \frac{x}{1+\sqrt{1+x^2}} - \sqrt{1+x^2}$$

$$= \sqrt{1+x^2}\,(\log x - 1) - \log \frac{x}{1+\sqrt{1+x^2}}$$

$$= \sqrt{1+x^2}\,\log \left(\frac{x}{e} \right) - \log \frac{x}{1+\sqrt{1+x^2}}.$$

$$(9.) \int e^x\,dx\,\frac{x^2+1}{(x+1)^2} = \int e^x\,dx\,\frac{x^2-1+2}{(x+1)^2}$$

$$= \int e^x\,dx \left\{ \frac{x-1}{x+1} + \frac{(x+1)-(x-1)}{(x+1)^2} \right\}$$

$$= \int e^x \left\{ \frac{(x-1)}{x+1}\,dx + d\left(\frac{x-1}{x+1} \right) \right\}$$

$$= e^x\,\frac{x-1}{x+1}.$$

(10.) $\displaystyle\int e^x \frac{(2-x^2)dx}{(1-x)\sqrt{1-x^2}} = \int e^x dx \frac{1-x^2+1}{(1-x)\sqrt{1-x^2}}$

$$= \int e^x dx \left\{ \frac{\sqrt{1+x}}{\sqrt{1-x}} + \frac{1}{(1-x)\sqrt{1-x^2}} \right\},$$

$$= \int e^x dx \left\{ \frac{\sqrt{1+x}}{\sqrt{1-x}} + \frac{(1-x)+(1+x)}{(1-x)^2} \frac{1}{2} \frac{\sqrt{1-x}}{\sqrt{1+x}} \right\}.$$

$$= \int e^x \left\{ \frac{\sqrt{1+x}}{\sqrt{1-x}} dx + d\left(\frac{\sqrt{1+x}}{\sqrt{1-x}} \right) \right\}$$

$$= e^x \frac{\sqrt{1+x}}{\sqrt{1-x}}.$$

(11.) $\displaystyle\int \frac{x e^x dx}{(e^x-1)^3} \qquad p = x \qquad dq = \frac{e^x dx}{(e^x-1)^3}$

$$dp = dx \qquad q = -\frac{dx}{2(e^x-1)^2},$$

$$\int \frac{x e^x dx}{(e^x-1)^3} = -\frac{x}{2(e^x-1)^2} + \frac{1}{2}\int \frac{dx}{(e^x-1)^2}$$

$$= -\frac{x}{2(e^x-1)^2} - \frac{1}{2}\int \frac{dx}{e^x-1} + \frac{1}{2}\int \frac{e^x dx}{(e^x-1)^2}$$

$$= -\frac{x}{2(e^x-1)^2} + \frac{1}{2}\int \frac{e^x dx}{e^x} - \frac{1}{2}\int \frac{e^x dx}{e^x-1} + \frac{1}{2}\int \frac{e^x dx}{(e^x-1)^2}$$

$$= -\frac{x}{2(e^x-1)^2} + \frac{1}{2} \log \frac{e^x}{e^x-1} - \frac{1}{2}\frac{1}{e^x-1}$$

$$= \frac{1}{2} \log \frac{e^x}{e^x-1} - \frac{1}{2(e^x-1)}\left\{ 1 + \frac{x}{e^x-1} \right\}.$$

(12.) $\dfrac{du}{dx} = \dfrac{a^x}{x^4}$ $p = a^x$ $\dfrac{dq}{dx} = \dfrac{1}{x^4}$,

$$\dfrac{dp}{dx} = \mathrm{A}\,a^x \qquad q = -\dfrac{1}{3\,x^3},$$

$$\therefore u = -\dfrac{a^x}{3\,x^3} + \dfrac{\mathrm{A}}{3}\int \dfrac{a^x\,dx}{x^3}.$$

Similarly,

$$\int \dfrac{a^x\,dx}{x^3} = -\dfrac{a^x}{2\,x^2} + \dfrac{\mathrm{A}}{2}\int \dfrac{a^x\,dx}{x^2}$$

$$\int \dfrac{a^x\,dx}{x^2} = -\dfrac{a^x}{x} + \mathrm{A}\int \dfrac{a^x\,dx}{x},$$

$$\therefore u = -\dfrac{a^x}{3\,x^3} + \dfrac{\mathrm{A}}{3}\left\{ -\dfrac{a^x}{2\,x^2} + \dfrac{\mathrm{A}}{2}\left(-\dfrac{a^x}{x} + \mathrm{A}\int \dfrac{a^x\,dx}{x} \right) \right\}$$

$$= -\,a^x \left\{ \dfrac{1}{3\,x^3} + \dfrac{\mathrm{A}}{2.3\,x^2} + \dfrac{\mathrm{A}^2}{2.3\,x} \right\} + \dfrac{\mathrm{A}^3}{2.3}\int \dfrac{a^x\,dx}{x}.$$

(13.) $\dfrac{du}{dx} = \dfrac{a^x}{\sqrt{x}}$ $p = x^{-\frac{1}{2}}$ $\dfrac{dq}{dx} = a^x$,

$$\dfrac{dp}{dx} = -\dfrac{1}{2}\,x^{-\frac{3}{2}} \qquad q = \dfrac{a^x}{\mathrm{A}},$$

$$\therefore u = \dfrac{a_x}{\mathrm{A}\,x^{\frac{1}{2}}} + \dfrac{1}{2\,\mathrm{A}}\int x^{-\frac{3}{2}}\,a^x\,dx,$$

also $\displaystyle\int x^{-\frac{3}{2}}\,a^x\,dx = \dfrac{a^x}{\mathrm{A}\,x^{\frac{3}{2}}} + \dfrac{3}{2\,\mathrm{A}}\int x^{-\frac{5}{2}}\,a^x\,dx$

$$\int x^{-\frac{5}{2}}\,a^x\,dx = \dfrac{a^x}{\mathrm{A}\,x^{\frac{5}{2}}} + \dfrac{5}{2\,\mathrm{A}}\int x^{-\frac{7}{2}}\,a^x\,dx,$$

E

$$\therefore u = \frac{a^x}{A\,x^{\frac{1}{2}}} + \frac{1}{2\,A}\left\{\frac{a^x}{A\,x^{\frac{3}{2}}} + \frac{3}{2\,A}\left(\frac{a^x}{A\,x^{\frac{5}{2}}} + \frac{5}{2\,A}\frac{a^x}{A\,x^{\frac{7}{2}}}\&c.\right)\right\}$$

$$= \frac{a^x}{A\,\sqrt{x}}\left\{1 + \frac{1}{2x\,A} + \frac{3}{(2\,x\,A)^2} + \frac{3\,.\,5}{(2\,x\,A)^3}\&c.\right\}.$$

Another result may be obtained by the following method.

$$\frac{d\,u}{d\,x} = \frac{a^x}{\sqrt{x}} \qquad p = a^x \qquad \frac{d\,q}{d\,x} = x^{-\frac{1}{2}}$$

$$\frac{d\,p}{d\,x} = A\,a^x \qquad q = 2\,x^{\frac{1}{2}},$$

$$\therefore u = 2\,a^x\,x^{\frac{1}{2}} - 2\,A\int a^x\,x^{\frac{1}{2}}\,dx.$$

Also, $\displaystyle\int a^x\,x^{\frac{1}{2}}\,dx = \frac{2}{3}\,a^x\,x^{\frac{3}{2}} - \frac{2}{3}\,A\int a^x\,x^{\frac{3}{2}}\,dx.$

$$\int a^x\,x^{\frac{3}{2}}\,dx = \frac{2}{5}\,a^x\,x^{\frac{5}{2}} - \frac{2}{5}\,A\int a^x\,x^{\frac{5}{2}}\,dx,$$

$$\int a^x\,x^{\frac{5}{2}}\,dx = \frac{2}{7}\,a^x\,x^{\frac{7}{2}} - \frac{2}{7}\,A\int a^x\,x^{\frac{7}{2}}\,dx,$$

$$\therefore u = 2\,a^x\,x^{\frac{1}{2}} - 2\,A\left\{\frac{2}{3}\,a^x\,x^{\frac{3}{2}} - \frac{2}{3}\,A\left(\frac{2}{5}\,a^x\,x^{\frac{5}{2}}\right.\right.$$

$$\left.\left. - \frac{2}{5}\,A\,\frac{2}{7}\,a^x\,x^{\frac{7}{2}}, \&c.\right)\right\},$$

$$\therefore u = \frac{a^x}{A\,\sqrt{x}}\left\{2\,x\,A - \frac{(2x\,A)^2}{3} + \frac{(2\,x\,A)^3}{3\,.\,5} - \frac{(2\,x\,A)^4}{3\,.\,5\,.\,7}\right.$$

$$\left. + \&c.\right\}.$$

$$(14.) \qquad \frac{d\,u}{d\,x} = x^n \, x \, x^m = x^m \left\{ 1 + n\,x \log x \right.$$

$$\left. + \frac{(n\,x \log x)^2}{1 \cdot 2} + \&\mathrm{c}. \right\}$$

$$= x^m + n\,x^{m+1} \log x + \frac{n^2}{2}\,x^{m+2}\,(\log x)^2 + \&\mathrm{c}.,$$

$$\therefore u = \frac{x^{m+1}}{m+1} + n \int x^{m+1} \log x \, dx$$

$$+ \frac{n^2}{1 \cdot 2} \int x^{m+2}\,(\log x)^2 \, dx + \&\mathrm{c}.$$

$$\text{But} \int x^m (\log x)^n \, dx = \frac{x^{m+1}}{m+1} \left\{ (\log x)^n - \frac{n}{m+1}(\log x)^{n-1} \right.$$

$$+ \frac{n\,(n-1)}{(m+1)^2}\,(\log x)^{n-2} - \&\mathrm{c}.$$

$$\left. \pm \frac{n\,(n-1)\,(n-2) \ldots 2 \cdot 1}{(m+1)^{n+1}}\,x^{m+1} \right\},$$

$$\therefore \int x^{m+1} \log x \, dx = \frac{x^{m+2}}{m+2} \left\{ \log x - \frac{1}{m+2} \right\}$$

$$\int x^{m+2}(\log x)^2 \, dx = \frac{x^{m+3}}{m+3} \left\{ (\log x)^2 \right.$$

$$\left. - \frac{2}{m+3} \log x + \frac{2}{(m+3)^2} \right\},$$

$$\int x^{m+3}\,(\log x)^3 \, dx = \frac{x^{m+4}}{m+4} \left\{ (\log x)^3 - \frac{3}{m+4}(\log x)^2 \right.$$

$$\left. + \frac{6}{(m+4)^2} \log x - \frac{6}{(m+4)^3} \right\}$$

Arranging the terms according to the powers of log x,

$$\int x^{nx} \cdot x^m \, dx = \frac{x^{m+1}}{m+1} - \frac{n\, x^{m+2}}{(m+2)^2} + \frac{n^2\, x^{m+3}}{(m+3)^3} - \frac{n^3\, x^{m+4}}{(m+4)^4} + \&c.$$

$$+ n \log x \left\{ \frac{x^{n+2}}{m+2} - \frac{n x^{m+3}}{(m+3)} + \frac{n^2 x^{m+4}}{(m+4)^3}, \&c. \right\}$$

$$+ \frac{n^2}{1 \cdot 2} (\log x)^2 \left\{ \frac{x^{m+3}}{m+3} - \frac{n\, x^{m+4}}{(m+4)^2} + \&c. \right\}.$$

If $x = 0$ all the terms vanish.

If $x = 1,$ $\int x^{nx} \cdot x^m \, dx$ becomes

$$\frac{1}{m+1} - \frac{n}{(m+2)^2} + \frac{n^2}{(m+3)^3} - \frac{n^3}{(m+4)^4} + \&c.$$

(15.) $\dfrac{d u}{d x} = \dfrac{\log x}{(1 + x)^4}$ $p = \log x$ $\dfrac{d q}{d x} = \dfrac{1}{(1 + x)^2}$

$$\frac{d p}{d x} = \frac{1}{x} \qquad q = - \frac{1}{1 + x},$$

$$\therefore u = - \frac{\log x}{1 + x} + \int \frac{d x}{x \, (1 + x)}$$

$$= - \frac{\log x}{1 + x} + \int \frac{d x}{x} - \int \frac{d x}{1 + x}$$

$$= - \frac{\log x}{1 + x} + \log x - \log (1 + x)$$

$$= \frac{x}{1 + x} \log x - \log (1 + x).$$

CHAPTER V.

(1.) $\int \sin^3 \theta \, d\theta = \int \sin \theta \, (1 - \cos^2 \theta) d\theta$

$$= \int \sin \theta \, d\theta - \int \sin \theta \cos^2 \theta \, d\theta$$

$$= - \cos \theta + \frac{\cos^3 \theta}{3} = - \cos \theta + \frac{\cos \theta \, (1 - \sin^2 \theta)}{3}$$

$$= - \frac{2}{3} \cos \theta - \frac{\cos \theta \sin^2 \theta}{3}$$

(2.) $\int \cos^3 \theta \, d\theta = \int \cos \theta \, (1 - \sin^2 \theta) \, d\theta$

$$= \int \cos \theta \, d\theta - \int \cos \theta \sin^2 \theta \, d\theta$$

$$= \sin \theta - \frac{\sin^3 \theta}{3} = \sin \theta - \frac{\sin \theta \, (1 - \cos^2 \theta)}{3}$$

$$= \frac{2 \sin \theta}{3} + \frac{\sin \theta \cos^2 \theta}{3}.$$

(3.) $\displaystyle\int \frac{d\theta}{\sin^3 \theta} = \int \frac{(\sin^2 \theta + \cos^2 \theta)\, d\theta}{\sin^3 \theta}$

$$= \int \frac{d\theta}{\sin \theta} + \int \frac{\cos^2 \theta\, d\theta}{\sin^3 \theta}$$

$$\int \cos \theta \, \frac{\cos\theta\, d\theta}{\sin^3 \theta} = -\cos \theta \, \frac{1}{2\sin^2 \theta} - \frac{1}{2}\int \frac{\sin\theta\, d\theta}{\sin^2\theta},$$

$$\therefore \int \frac{d\theta}{\sin^3 \theta} = \frac{1}{2}\int \frac{d\theta}{\sin \theta} - \frac{1\cos\theta}{2\sin^2\theta}$$

$$= \tfrac{1}{2}\log\left(\tan\frac{\theta}{2}\right) - \frac{1\cos\theta}{2\sin^2\theta}.$$

(4.) $\displaystyle\int \frac{\sin^3 \theta\, d\theta}{\cos^4 \theta} = \int \frac{\sin \theta\, d\theta}{\cos^4 \theta} - \int \frac{\sin \theta\, d\theta}{\cos^2 \theta}$

$$= \frac{1}{3\cos^3 \theta} - \frac{1}{\cos \theta} = \frac{1}{3\cos^3 \theta}\left\{\frac{1}{3} - 1 + \sin^2 \theta\right\}$$

$$= \frac{1}{3\cos^3 \theta}\left\{\sin^2 \theta - \frac{2}{3}\right\}.$$

(5.) $\displaystyle\int x^2 \tan^{-1}\sqrt{\frac{x}{a}}\, dx.$ Let $\dfrac{x}{a} = z^2$

$$x^3 = a^3 z^6 \qquad\qquad x^2\, dx = 2\,a^3\, z^5\, dz,$$

$$\therefore \int x^2 \tan^{-1}\sqrt{\frac{x}{a}}\, dx = 2\,a^3 \int z^5 \tan^{-1} z\, dz$$

$$p = \tan^{-1} z \qquad\qquad dq = 2\,a^3\, z^5\, dz$$

$$dp = \frac{1}{1 + z^2}\, dz \qquad\qquad q = \frac{a^3 z^6}{3},$$

$$\therefore \ u = \frac{a^3 z^6}{3} \tan^{-1} z - \frac{a^3}{3} \int \frac{z^5}{1+z^2} dz$$

$$= \frac{a^3 z^6}{3} \tan^{-1} z - \frac{a^3}{3} \int \left\{ z^4 - z^2 + 1 - \frac{1}{1+z^2} \right\} dz$$

$$= \frac{a^3 z^6}{3} \tan^{-1} z - \frac{a^3}{3} \left\{ \frac{z^5}{5} - \frac{z^3}{3} + z \right\} + \frac{a^3}{3} \tan^{-1} z$$

$$= \frac{a^3}{3} (z^6 + 1) \tan^{-1} z - \frac{a^3 z}{3} \left\{ \frac{z^4}{5} - \frac{z^2}{3} + 1 \right\}$$

$$= \frac{a^3}{3} \left(\frac{x^3 + a^3}{a^3} \right) \tan^{-1} \sqrt{\frac{x}{a}} - \frac{a^3 \sqrt{x}}{3 \sqrt{a}} \left\{ \frac{x^2}{5 a^2} - \frac{x}{3 a} + 1 \right\}$$

$$= \frac{x^3 + a^3}{3} \tan^{-1} \sqrt{\frac{x}{a}} - \frac{\sqrt{a x}}{3} \left(\frac{x^2}{5} - \frac{a x}{3} + a^2 \right).$$

(6) $\dfrac{du}{d\theta} = \cos^6 \theta \qquad p = \cos^5 \theta \qquad \dfrac{dq}{d\theta} = \cos \theta$

$$\frac{dp}{d\theta} = - 5 \cos^4 \theta \sin \theta \quad q = \sin \theta,$$

$$\therefore \int \cos^6 \theta \, d\theta = \sin \theta \cos^5 \theta + 5 \int \cos^4 \theta \sin^2 \theta \, d\theta$$

$$= \sin \theta \cos^5 \theta + 5 \int \cos^4 \theta \, dx - 5 \int \cos^6 \theta \, dx,$$

$$\therefore \int \cos^6 \theta \, d\theta = \frac{\sin \theta \cos^5 \theta}{6} + \frac{5}{6} \int \cos^4 \theta \, d\theta.$$

Similarly,

$$\int \cos^4 \theta \, d\theta = \frac{1}{4} \sin \theta \cos^3 \theta + \frac{3}{4} \int \cos^2 \theta \, d\theta$$

$$\int \cos^2 \theta \, d\theta = \frac{1}{2} \sin \theta \, \cos \theta + \frac{1}{2} \int d\theta = \frac{1}{2} \sin \theta \cos \theta + \frac{\theta}{2},$$

$$\therefore \int \cos^6 \theta \, d\theta = \frac{\sin \theta \cos^5 \theta}{6} + \frac{5}{6} \left\{ \frac{1}{4} \sin \theta \cos^3 \theta + \frac{3}{4} \right.$$

$$\left. \left(\frac{1}{2} \sin \theta \cos \theta + \frac{\theta}{2} \right) \right\}$$

$$= \sin \theta \left\{ \frac{\cos^5 \theta}{6} + \frac{5 \cos^3 \theta}{24} + \frac{5 \cos \theta}{16} \right\} + \frac{5 \theta}{16}.$$

(7.) $\dfrac{du}{d\theta} = \sin^2 \theta \cos^4 \theta = \cos^4 \theta - \cos^6 \theta.$

But from preceding example,

$$\int \cos^4 \theta \, d\theta = \sin \theta \left\{ \frac{\cos^3 \theta}{4} + \frac{3 \cos \theta}{8} \right\} + \frac{3 \theta}{8}$$

$$\int \cos^6 \theta \, d\theta = \sin \theta \left\{ \frac{\cos^5 \theta}{6} + \frac{5 \cos^3 \theta}{24} + \frac{5 \cos \theta}{16} \right\} + \frac{5 \theta}{16}$$

$$\therefore u = \sin \theta \left\{ - \frac{\cos^5 \theta}{6} + \frac{\cos^3 \theta}{24} + \frac{\cos \theta}{16} \right\} + \frac{\theta}{16}$$

$$= \sin \theta \left\{ - \frac{\cos^5 \theta}{6} + \frac{\sin^2 \theta \cos^3 \theta}{6} + \frac{\cos^3 \theta}{24} + \frac{\cos \theta}{16} \right\} + \frac{\theta}{16}$$

$$= \sin \theta \left\{ \frac{\sin^2 \theta \cos^3 \theta}{6} - \frac{1}{8} \cos \theta \left(1 - \sin^2 \theta \right) + \frac{\cos \theta}{16} \right\} + \frac{\theta}{16}$$

$$= \frac{\sin^3 \theta \cos^3 \theta}{6} + \frac{\sin^3 \theta \cos \theta}{8} - \frac{\sin \theta \cos \theta}{16} + \frac{\theta}{16}.$$

(8.) $\dfrac{du}{d\theta} = \sin^6\theta\cos^3\theta = \sin^6\theta\cos\theta - \sin^8\theta\cos\theta,$

$$u = \frac{\sin^7\theta}{7} - \frac{\sin^9\theta}{9} = \sin^7\theta\left\{\frac{1}{7} - \frac{\sin^2\theta}{9}\right\}$$

$$= \sin^7\theta\left\{\frac{1}{7} - \frac{1-\cos^2\theta}{9}\right\}$$

$$= \sin^7\theta\left\{\frac{\cos^2\theta}{9} + \frac{2}{63}\right\}.$$

(9.) $\displaystyle\int\frac{d\theta}{\sin^5\theta} = \int\frac{\cos^2\theta\, d\theta}{\sin^5\theta} + \int\frac{d\theta}{\sin^3\theta},$

$$\int\frac{\cos^2\theta\, d\theta}{\sin^5\theta} \begin{cases} p = \cos\theta & \dfrac{dq}{d\theta} = \dfrac{\cos\theta}{\sin^5\theta} \\[2mm] \dfrac{dp}{d\theta} = -\sin\theta & q = -\dfrac{1}{4\sin^4\theta} \end{cases}$$

$$\int\frac{\cos^2\theta\, d\theta}{\sin^5\theta} = -\frac{\cos\theta}{4\sin^4\theta} - \frac{1}{4}\int\frac{d\theta}{\sin^3\theta},$$

$$\therefore\; u = -\frac{\cos\theta}{4\sin^4\theta} + \frac{3}{4}\int\frac{d\theta}{\sin^3\theta},$$

$$\int\frac{d\theta}{\sin^3\theta} = -\frac{\cos\theta}{2\sin^2\theta} + \frac{1}{2}\int\frac{d\theta}{\sin\theta},$$

$$\int\frac{d\theta}{\sin\theta} = -\frac{\cos\theta}{2\sin^2\theta} + \frac{1}{2}\log\left(\tan\frac{\theta}{2}\right),$$

$$\therefore\; u = -\frac{\cos\theta}{4\sin^4\theta} - \frac{3\cos\theta}{8\sin^2\theta} + \frac{3}{8}\log\left(\tan\frac{\theta}{2}\right)$$

$$= -\cos\theta\left\{\frac{1}{4\sin^4\theta} + \frac{3}{8\sin^2\theta}\right\} + \frac{3}{8}\log\left(\tan\frac{\theta}{2}\right).$$

E 3

(10.) $\displaystyle \int \frac{d\theta}{\cos^6 \theta} = \int \frac{\sin^2 \theta \, d\theta}{\cos^6 \theta} + \int \frac{d\theta}{\cos^4 \theta}$

$$p = \sin \theta \qquad \frac{dq}{d\theta} = \frac{\sin \theta}{\cos^6 \theta},$$

$$\frac{dp}{d\theta} = \cos \theta \qquad q = \frac{1}{5 \cos^5 \theta},$$

$$\therefore \int \frac{\sin^2 \theta \, d\theta}{\cos^6 \theta} = \frac{\sin \theta}{5 \cos^5 \theta} - \frac{1}{5} \int \frac{d\theta}{\cos^4 \theta},$$

$$\therefore u = \frac{\sin \theta}{5 \cos^5 \theta} + \frac{4}{5} \int \frac{d\theta}{\cos^4 \theta}.$$

Similarly,

$$\int \frac{d\theta}{\cos^4 \theta} = \frac{\sin \theta}{3 \cos^3 \theta} + \frac{2}{3} \int \frac{d\theta}{\cos^2 \theta}$$

$$= \frac{\sin \theta}{3 \cos^3 \theta} + \frac{2}{3} \frac{\sin \theta}{\cos \theta},$$

$$\therefore u = \frac{\sin \theta}{5 \cos^5 \theta} + \frac{4}{5} \left\{ \frac{\sin \theta}{3 \cos^3 \theta} + \frac{2 \sin \theta}{3 \cos \theta} \right\}$$

$$= \sin \theta \left\{ \frac{1}{5 \cos^5 \theta} + \frac{4}{15 \cos^3 \theta} + \frac{8}{15 \cos \theta} \right\}.$$

(11.) $\displaystyle \frac{du}{d\theta} = \frac{\sin^5 \theta}{\cos^2 \theta} = \frac{\sin \theta (1 - 2 \cos^2 \theta + \cos^4 \theta)}{\cos^2 \theta}$

$$= \frac{\sin \theta}{\cos^2 \theta} - 2 \sin \theta + \sin \theta \cos^2 \theta,$$

$$\therefore u = \frac{1}{\cos \theta} + 2 \cos \theta - \frac{\cos^3 \theta}{3}$$

$$= \frac{1}{\cos\theta}\left\{1 + 2\cos^2\theta - \frac{\cos^4\theta}{3}\right\}$$

$$= \frac{1}{\cos\theta}\left\{1 + 2 - 2\sin^2\theta - \frac{1 - 2\sin^2\theta + \sin^4\theta}{3}\right\}$$

$$= -\frac{1}{\cos\theta}\left\{\frac{\sin^4\theta}{3} + \frac{4\sin^2\theta}{3} - \frac{8}{3}\right\}.$$

(12.) $\dfrac{du}{d\theta} = \dfrac{\cos^4\theta}{\sin^3\theta}$ $p = \cos^3\theta$ $\dfrac{dq}{d\theta} = \dfrac{\cos\theta}{\sin^3\theta}$

$$\frac{dp}{d\theta} = -3\sin\theta\cos^2\theta \quad q = -\frac{1}{2\sin^2}$$

$$\therefore u = -\frac{\cos^3\theta}{2\sin^2\theta} - \frac{3}{2}\int\frac{\cos^2\theta\, d\theta}{\sin\theta}$$

$$= -\frac{\cos^3\theta}{2\sin^2\theta} - \frac{3}{2}\int\frac{\cos^2\theta\, d\theta}{\sin^3\theta} + \frac{3}{2}\int\frac{\cos^4\theta\, d\theta}{\sin^3\theta},$$

$$\therefore u = \frac{\cos^3\theta}{\sin^2\theta} + 3\int\frac{\cos^2\theta\, d\theta}{\sin^3\theta}$$

$$p = \cos\theta \qquad \frac{dq}{d\theta} = \frac{\cos\theta}{\sin^3\theta}$$

$$\frac{dp}{d\theta} = -\sin\theta \qquad q = -\frac{1}{2\sin^2\theta}$$

$$\int\frac{\cos^2\theta\, d\theta}{\sin^3\theta} = -\frac{\cos\theta}{2\sin^2\theta} - \frac{1}{2}\int\frac{d\theta}{\sin\theta}$$

$$= -\frac{\cos\theta}{2\sin^2\theta} - \frac{1}{2}\log\left(\tan\frac{\theta}{2}\right),$$

$$\therefore\; u = \frac{\cos^3\theta}{\sin^2\theta} - \frac{3\cos\theta}{2\sin^2\theta} - \frac{3}{2}\log\left(\tan\frac{\theta}{2}\right)$$

$$= \frac{1}{\sin^2\theta}\left\{\cos^3\theta - \frac{3\cos\theta}{2}\right\} - \frac{3}{2}\log\left(\tan\frac{\theta}{2}\right)$$

(13.)
$$\frac{du}{d\theta} = \frac{1}{\sin^2\theta\cos^3\theta} = \frac{1}{\cos^3\theta} + \frac{1}{\sin^2\theta\cos\theta}$$

$$= \frac{\sin^2\theta}{\cos^3\theta} + \frac{1}{\cos\theta} + \frac{1}{\cos\theta} + \frac{\cos\theta}{\sin^2\theta},$$

$$u = \int \frac{\sin^2\theta\, d\theta}{\cos^3\theta} + 2\int \frac{d\theta}{\cos\theta} - \frac{1}{\sin\theta}$$

$$p = \sin\theta \qquad\qquad \frac{dq}{d\theta} = \frac{\sin\theta}{\cos^3\theta}$$

$$\frac{dp}{d\theta} = \cos\theta \qquad\qquad q = \frac{1}{2\cos^2\theta},$$

$$\therefore \int \frac{\sin^2\theta\, d\theta}{\cos^3\theta} = \frac{\sin\theta}{2\cos^2\theta} - \frac{1}{2}\int \frac{d\theta}{\cos\theta},$$

$$.\; u = \frac{\sin\theta}{2\cos^2\theta} + \frac{3}{2}\int \frac{d\theta}{\cos\theta} - \frac{1}{\sin\theta}$$

$$= \frac{1}{\sin\theta}\left\{\frac{1-\cos^2\theta}{2\cos^2\theta} - 1\right\} + \frac{3}{2}\log\left\{\tan\left(\frac{\pi}{4} + \frac{\theta}{2}\right)\right\}$$

$$= \frac{1}{\sin\theta}\left\{\frac{1}{2\cos^2\theta} - \frac{3}{2}\right\} + \frac{3}{2}\log\tan\left(\frac{\pi}{4} + \frac{\theta}{2}\right).$$

$$(14.) \quad \frac{du}{d\theta} = \frac{1}{\sin^4\theta \cos^2\theta} = \frac{1}{\sin^2\theta \cos^2\theta} + \frac{1}{\sin^4\theta}$$

$$= \frac{1}{\cos^2\theta} + \frac{1}{\sin^2\theta} + \frac{1}{\sin^4\theta} + \frac{\cos^2\theta}{\sin^4\theta},$$

$$\therefore \quad u = \tan\theta + 2\int\frac{d\theta}{\sin^2\theta} + \int\frac{\cos^2\theta\, d\theta}{\sin^4\theta}$$

$$\int\frac{\cos^2\theta\, d\theta}{\sin^4\theta} \begin{cases} p = \cos\theta & \dfrac{dq}{d\theta} = \dfrac{\cos\theta}{\sin^4\theta} \\[2ex] \dfrac{dp}{d\theta} = -\sin\theta & q = -\dfrac{1}{3\sin^3\theta} \end{cases}$$

$$= -\frac{\cos\theta}{3\sin^3\theta} - \frac{1}{3}\int\frac{d\theta}{\sin^2\theta},$$

$$\therefore \quad u = -\frac{\cos\theta}{3\sin^3\theta} + \tan\theta + \frac{5}{3}\int\frac{d\theta}{\sin^2\theta}$$

$$= -\frac{\cos^2\theta}{3\sin^3\theta\cos\theta} + \frac{\sin\theta}{\cos\theta} - \frac{5}{3}\frac{\cos\theta}{\sin\theta}$$

$$= -\frac{1-\sin^2\theta}{3\sin^3\theta\cos\theta} + \frac{3\sin^2\theta - 5\cos^2\theta}{3\sin\theta\cos\theta}$$

$$= -\frac{1}{3\sin^3\theta\cos\theta} + \frac{1 + 3\sin^2\theta - 5\cos^2\theta}{3\sin\theta\cos\theta}.$$

$$= -\frac{1}{3\sin^3\theta\cos\theta} + \frac{4(\sin^2\theta - \cos^2\theta)}{3\sin\theta\cos\theta}$$

$$= -\frac{1}{3\sin^3\theta\cos\theta} - \frac{8}{3}\frac{\cos^2\theta - \sin^2\theta}{2\sin\theta\cos\theta}$$

$$= -\frac{1}{3\sin^3\theta\cos\theta} - \frac{8}{3}\cot 2\theta.$$

(15.) $\dfrac{du}{d\theta} = \tan^4\theta = \tan^2\theta\,(1 + \tan^2\theta) - \tan^2\theta$

$$= \tan^2\theta\,(1 + \tan^2\theta) - (1 + \tan^2\theta) + 1,$$

$\therefore\ u = \dfrac{\tan^3\theta}{3} - \tan\theta + \theta.$

(16.) $\dfrac{du}{d\theta} = \dfrac{1}{\tan^5\theta} = \dfrac{1 + \tan^2\theta}{\tan^5\theta} - \dfrac{1}{\tan^3\theta}$

$$= \dfrac{1 + \tan^2\theta}{\tan^5\theta} - \dfrac{1 + \tan^2\theta}{\tan^3\theta} + \dfrac{1}{\tan\theta},$$

$\therefore\ u = \dfrac{-1}{4\tan^4\theta} + \dfrac{1}{2\tan^2\theta} + \text{h.l.}\sin\theta.$

(17.) $\dfrac{du}{d\theta} = \theta^3\cos\theta \qquad p = \theta^3 \qquad \dfrac{dq}{d\theta} = \cos\theta$

$$\dfrac{dp}{d\theta} = 3\theta^2 \qquad q = \sin\theta,$$

$$\therefore\ u = \theta^3\sin\theta - 3\int\theta^2\sin\theta\,d\theta$$

$$p = \theta^2 \qquad \dfrac{dq}{d\theta} = \sin\theta$$

$$\dfrac{dp}{d\theta} = 2\theta \qquad q = -\cos\theta$$

$$\int\theta^2\sin\theta\,d\theta = -\theta^2\cos\theta + 2\int\theta\cos\theta\,d\theta$$

$$p = \theta \qquad \dfrac{dq}{d\theta} = \cos\theta$$

$$\dfrac{dp}{d\theta} = 1 \qquad q = \sin\theta$$

$$\int \theta \cos \theta \, d\theta = \theta \sin \theta - \int \sin \theta \, d\theta = \theta \sin \theta + \cos \theta,$$

$$\therefore u = \theta^3 \sin \theta + 3\theta^2 \cos \theta - 6\theta \sin \theta - 6 \cos \theta.$$

(18.) $\dfrac{du}{dx} = \dfrac{x^2}{\sqrt{1-x^2}} \sin^{-1} x.$

To integrate $\dfrac{x^2 dx}{\sqrt{1-x^2}}$ $p = x$ $\dfrac{dq}{dx} = \dfrac{x}{\sqrt{1-x^2}}$

$$\dfrac{dp}{dx} = 1 \qquad q = -\sqrt{1-x^2}$$

$$\int \frac{x^2 \, dx}{\sqrt{1-x^2}} = -x\sqrt{1-x^2} + \int \sqrt{1-x^2} \, dx$$

$$= -x\sqrt{1-x^2} + \int \frac{dx}{\sqrt{1-x^2}} - \int \frac{x^2 \, dx}{\sqrt{1-x^2}},$$

$$\therefore \int \frac{x^2 \, dx}{\sqrt{1-x^2}} = -\frac{x}{2}\sqrt{1-x^2} + \frac{1}{2}\sin^{-1} x$$

Next, to integrate $\dfrac{x^2 \, dx}{\sqrt{1-x^2}} \sin^{-1} x,$

$$p = \sin^{-1} x \qquad \frac{dq}{dx} = \frac{x^2}{\sqrt{1-x^2}}$$

$$\frac{dp}{dx} = \frac{1}{\sqrt{1-x^2}} \qquad q = -\frac{x}{2}\sqrt{1-x^2} + \frac{1}{2}\sin^{-1} x,$$

$$\therefore u = \frac{1}{2}(\sin^{-1} x)^2 - \frac{x}{2}\sqrt{1-x^2}\sin^{-1} x + \frac{1}{2}\int x \, dx$$

$$-\frac{1}{2}\int \frac{\sin^{-1} x \, dx}{\sqrt{1-x^2}}$$

$$= \frac{1}{2}(\sin^{-1} x)^2 - \frac{x\sqrt{1-x^2}}{2}\sin^{-1} x + \frac{x^2}{4} - \frac{1}{4}(\sin^{-1} x)^2$$

$$= \frac{1}{4}(\sin^{-1} x)^2 - \frac{x\sqrt{1-x^2}}{2}\sin^{-1} x + \frac{x^2}{4}.$$

(19.) $\dfrac{du}{dx} = \dfrac{x}{(1-x^2)^{\frac{3}{2}}}\sin^{-1} x.$

$$p = \sin^{-1} x \qquad \frac{dq}{dx} = \frac{x}{(1-x^2)^{\frac{3}{2}}}$$

$$\frac{dp}{dx} = \frac{1}{\sqrt{1-x^2}} \qquad q = \frac{1}{(1-x^2)^{\frac{1}{2}}},$$

$$\therefore u = \frac{\sin^{-1} x}{\sqrt{1-x^2}} - \int \frac{dx}{1-x^2}$$

$$= \frac{\sin^{-1} x}{\sqrt{1-x^2}} - \frac{1}{2}\left\{ \int \frac{dx}{1-x} + \int \frac{dx}{1+x} \right\}$$

$$= \frac{\sin^{-1} x}{\sqrt{1-x^2}} + \log\frac{\sqrt{1-x}}{\sqrt{1+x}}.$$

(20.) $\dfrac{du}{dx} = \dfrac{x^2}{1+x^2}\tan^{-1} x.$

$$p = \tan^{-1} x \qquad \frac{dq}{dx} = \frac{x^2}{1+x^2} = 1 - \frac{1}{1+x^2}$$

$$\frac{dp}{dx} = \frac{1}{1+x^2} \qquad q = x - \tan^{-1} x,$$

$$\therefore u = \tan^{-1} x\,(x - \tan^{-1} x) - \int \frac{x\,dx}{1+x^2} + \int \frac{\tan^{-1} x\,dx}{1+x^2}.$$

$$= \tan^{-1} x \, (x - \tan^{-1} x) - \log \sqrt{1 + x^2} + \frac{(\tan^{-1} x)^2}{2}$$

$$= x \tan^{-1} x - \frac{1}{2} (\tan^{-1} x)^2 - \log \sqrt{1 + x^2}.$$

(21.) $\dfrac{du}{dx} = e^{ax} \sin^2 x.$

$$p = \sin^2 x \qquad \frac{dq}{dx} = e^{ax}$$

$$\frac{dp}{dx} = 2 \sin x \cos x \quad q = \frac{e^{ax}}{a}$$

$$u = \frac{1}{a} e^{ax} \sin^2 x - \frac{2}{a} \int e^{ax} \sin x \cos x \, dx$$

$$p = \sin x \cos x \qquad \frac{dq}{dx} = e^{ax}$$

$$\frac{dp}{dx} = \cos^2 x - \sin^2 x \qquad q = \frac{e^{ax}}{a}$$

$$= 1 - 2 \sin^2 x ,$$

$$\therefore \int e^{ax} \sin x \cos x \, dx$$

$$= \frac{1}{a} e^{ax} \sin x \cos x - \frac{1}{a} \int e^{ax} \, dx + \frac{2}{a} \int e^{ax} \sin^2 x \, dx,$$

$$\therefore u = \frac{1}{a} e^{ax} \sin^2 x - \frac{2}{a^2} e^{ax} \sin x \cos x + \frac{2}{a^2} \frac{e^{ax}}{a} - \frac{4u}{a^2},$$

$$\therefore u \left(1 + \frac{4}{a^2} \right) = \frac{e^{ax} \sin x}{a^2} (a \sin x - 2 \cos x) + \frac{2 e^{ax}}{a^3},$$

$$\therefore u = \frac{e^{ax} \sin x \, (a \sin x - 2 \cos x)}{a^2 + 4} + \frac{2 e^{ax}}{a (a^2 + 4)}.$$

(22.) $\dfrac{du}{dx} = \dfrac{1}{(a + b \cos x)^2}.$

Assume

$$\int \frac{dx}{(a + b \cos x)^2} = \frac{A \sin x}{a + b \cos x} + B \int \frac{dx}{a + b \cos x}.$$

Taking the differential coefficients

$$\frac{1}{(a + b \cos x)^2} = \frac{A \cos x (a + b \cos x) + b A \sin^2 x}{(a + b \cos x)^2}$$

$$+ \frac{B (a + b \cos x)}{(a + b \cos x)^2},$$

$$\therefore \; 1 = A \cos x (a + b \cos x) + b A \sin^2 x + B (a + b \cos x)$$
$$= A a \cos x + A b \cos^2 x + A b \sin^2 x + B a + B b \cos x$$
$$= (A a + B b) \cos x + A b + B a.$$

Equating like powers of $(\cos x)$,

$$A a + B b = 0, \qquad \therefore \; A = - \frac{b}{a} B$$

$$A b + B a = 1$$

$$B \left(- \frac{b^2}{a} + a \right) = 1, \qquad \therefore \; B = \frac{a}{a^2 - b^2},$$

$$\therefore \; A = - \frac{b}{a^2 - b^2},$$

$$\therefore \int \frac{dx}{(a + b \cos x)^2}$$

$$= - \frac{b}{a^2 - b^2} \frac{\sin x}{a + b \cos x} + \frac{a}{a^2 - b^2} \int \frac{dx}{a + b \cos x}$$

$$= \frac{1}{a^2 - b^2} \left\{ \frac{- b \sin x}{a + b \cos x} + a \int \frac{dx}{a + b \cos x} \right\}.$$

(23.) To integrate $e^{ax} \cos k x$.

$$\text{Let } p = \cos k x \qquad dq = e^{ax} dx,$$

$$dp = - k \sin kx \, dx \qquad q = \frac{e^{ax}}{a},$$

(1) $\therefore \int e^{ax} \cos kx dx = \frac{e^{ax} \cos kx}{a} + \frac{k}{a} \int e^{ax} \sin kx dx.$

(2) $\int e^{ax} \sin kx \, dx = \frac{e^{ax} \sin kx}{a} - \frac{k}{a} \int e^{ax} \cos kx dx.$

Multiplying equation (2) by $\dfrac{k}{a}$ and substituting in equation (1)

$$\int e^{ax} \cos kx dx \left(1 + \frac{k^2}{a^2} \right) = \frac{e^{ax} \cos kx}{a} + \frac{k e^{ax} \sin kx}{a^2},$$

$$\int e^{ax} \cos kx dx = \frac{e^{ax} (a \cos kx + k \sin kx)}{a^2 + k^2}.$$

(24.) To integrate $e^{-ax} \sin k x$,

$$p = \sin k x \qquad d q = e^{-ax} dx,$$

$$dp = k \cos k x \qquad q = - \frac{e^{-ax}}{a},$$

(1) $\therefore \int e^{-ax} \sin kx \, dx = - \frac{e^{-ax} \sin kx}{a} + \frac{k}{a} \int e^{-ax} \cos kx \, dx.$

(2) $\int e^{-ax} \cos kx \, dx$

$$= - \frac{e^{-ax} \cos kx}{a} - \frac{k}{a} \int e^{-ax} \sin kx \, dx.$$

Multiplying equation (2) by $\left(\dfrac{k}{a}\right)$ and substituting in equation (1)

$$\int e^{-ax} \sin kx\, dx = -\frac{e^{-ax}\sin kx}{a} - \frac{ke^{-ax}\cos kx}{a^2}$$

$$-\frac{k^2}{a^2}\int e^{ax}\sin kx\, dx$$

$$\frac{a^2 + k^2}{a^2}\int e^{-ax}\sin kx\, dx = -\frac{e^{-ax}(a\sin kx + k\cos kx)}{a^2}$$

$$\int e^{-ax}\sin kx\, dx = -\frac{e^{-ax}(a\sin kx + k\cos kx)}{a^2 + k^2}.$$

(25.) $$\int \frac{d\theta}{a\cos^2\theta + b\sin^2\theta} = \int \frac{\dfrac{d\theta}{\cos^2\theta}}{a + b\tan^2\theta}$$

$$= \int \frac{d(\tan\theta)}{a + b\tan^2\theta} = \frac{1}{\sqrt{ab}}\tan^{-1}\left(\tan\theta\sqrt{\frac{b}{a}}\right).$$

(26.) $$\int \frac{\cos\theta\, d\theta}{(1 - e^2\cos^2\theta)^{\frac{3}{2}}} = \int \frac{\cos\theta\, d\theta}{(1 - e^2 + e^2\sin^2\theta)^{\frac{3}{2}}}$$

$$= \int \frac{\dfrac{\cos\theta}{\sin^3\theta}\, d\theta}{(1 - e^2)^{\frac{3}{2}}\left(\dfrac{1}{\sin^2\theta} + \dfrac{e^2}{1 - e^2}\right)^{\frac{3}{2}}}$$

$$= -\int \frac{\dfrac{1}{2}\, d\left(\dfrac{1}{\sin^2\theta}\right)}{(1 - e^2)^{\frac{3}{2}}\left(\dfrac{1}{\sin^2\theta} + \dfrac{e^2}{1 - e^2}\right)^{\frac{3}{2}}}$$

$$= \frac{1}{(1-e^2)^{\frac{3}{2}} \sqrt{\dfrac{1}{\sin^2 \theta} + \dfrac{e^2}{1-e^2}}}$$

$$= \frac{\sin \theta}{(1-e^2) \sqrt{1 - e^2 + e^2 \sin^2 \theta}}$$

$$= \frac{\sin \theta}{(1-e^2)\sqrt{1 - e^2 \cos^2 \theta}}$$

(27.) $\displaystyle\int (\cos 2\theta)^{\frac{3}{2}} \cos \theta \, d\theta = \int (1 - 2 \sin^2 \theta)^{\frac{3}{2}} \cos \theta \, d\theta.$

Let $2 \sin^2 \theta = x^2 \qquad d\theta = \dfrac{dx}{\sqrt{2} \cos \theta}$

$$du = \frac{(1-x^2)^{\frac{3}{2}} dx}{\sqrt{2}} = \frac{(1 - 2x^2 + x^4)\, dx}{\sqrt{2}\sqrt{1-x^2}}$$

$$\sqrt{2}\, du = \frac{dx}{\sqrt{1-x^2}} - \frac{2x^2\, dx}{\sqrt{1-x^2}} + \frac{x^4\, dx}{\sqrt{1-x^2}}$$

$$\int \frac{x^2}{\sqrt{1-x^2}}\, dx = -\frac{x}{2}\sqrt{1-x^2} + \frac{1}{2}\sin^- \; x,$$

$$\int \frac{x^4}{\sqrt{1-x^2}}\, dx = -\frac{x^3}{4}\sqrt{1-x^2} + \frac{3}{4}\int \frac{x^2}{\sqrt{1-x^2}}\, dx$$

$$= -\sqrt{1-x^2}\left(\frac{x^3}{4} + \frac{3x}{2 \cdot 4}\right) + \frac{3}{8}\sin^{-1} x,$$

$$\therefore \; \sqrt{2}\, u = \sin^{-1} x + x\sqrt{1-x^2} - \sin^{-1} x$$

$$- \sqrt{1-x^2}\left(\frac{x^3}{4} + \frac{3x}{8}\right) + \frac{3}{8}\sin^{-1} x$$

$$= \sqrt{1-x^2} \left(\frac{5x}{8} - \frac{x^3}{4} \right) + \frac{3}{8} \sin^{-1} x$$

$$\sqrt{2}u = \sqrt{\cos 2\theta} \left(\frac{5\sqrt{2}\sin\theta}{8} - \frac{2\sqrt{2}\sin^3\theta}{4} \right)$$

$$+ \frac{3}{8} \sin^{-1} (\sqrt{2} \sin \theta),$$

$$u = \sqrt{\cos 2\theta} \; \frac{\sin\theta}{8} (5 - 4 + 4 \cos^2 \theta)$$

$$+ \frac{3}{8\sqrt{2}} \sin^{-1} (\sqrt{2} \sin \theta)$$

$$= \frac{\sin\theta}{8} (3 + 2\cos 2\theta) \sqrt{\cos 2\theta} + \frac{3}{8\sqrt{2}} \sin^{-1} (\sqrt{2} \sin \theta).$$

(28.) $$\int \frac{\sin\theta \, d\theta}{\sqrt{\sin^2 a - \sin^2 \theta}} = \int \frac{\sin\theta \, d\theta}{\sqrt{\cos^2\theta - \cos^2 a}}$$

$$= - \int \frac{d(\cos\theta)}{\sqrt{\cos^2\theta - \cos^2 a}} = - \log \left(\sqrt{\cos^2\theta - \cos^2 a} + \cos\theta \right)$$

$$= - \log \left(\sqrt{\sin^2 a - \sin^2 \theta} + \cos\theta \right) + C.$$

If $\theta = a$, $u = - \log (\cos a) + C$,

$$\therefore \int_a^\theta \frac{\sin\theta \, d\theta}{\sqrt{\sin^2 a - \sin^2 \theta}} = \log \frac{\cos a}{\sqrt{\sin^2 a - \sin^2 \theta} + \cos\theta}.$$

CHAPTER VI

ON DEFINITE INTEGRALS.

In the process of differentiation all constant quantities which are merely added to, or subtracted from, those quantities which contain the variable disappear; and, on the contrary, after integration, there may be a constant quantity connected with the integral which we have not in that operation obtained. The letter C is therefore added to every integral to represent this quantity, and in order to determine its value we must in the first place find what particular value of the variable makes the integral 0; we thus get two equations, both of which contain C, and between these two equations C may be eliminated. For instance, if we have $\int 3 x^2 d4$ $= x^3 + C$ for the *general value* of the integral, and the problem indicates that for the *particular value* $x = a$, the integral becomes 0, then $0 = a^3 + C$; by subtracting this from the general value of the integral we get $x^3 - a^3$, in which the constant C has disappeared; this latter is called the *corrected integral*, and is written thus,

$$\int_a^x 3 x^2 d x = x^3 - a^3.$$

In this expression the value of the integral commences when $x = a$, and if we give another value to x, say $x = b$, then we have fully determined the value of the integral, which is now written

$$\int_a^b 3 x^2 d x = b^3 - a^3.$$

This is called the *definite integral*, and is said to be taken between the limits $x = b$ and $x = a$: the former is called the superior limit, and the latter the inferior limit, and the operation is called integration between limits.

$$\text{In general} \int f(x)\,dx = \phi(x) + C.$$

In order to determine the definite integral, we must, according to the nature of the problem proposed, assign the proper limits, which we shall represent by a and b, as before; then we have

$$\int_a^b f(x)\,dx = \phi(a) - \phi(b).$$

As every function of x may represent the ordinates of a curve whose abscissa is x, it follows that the operation of integrating between limits may be applied to finding the areas and lengths of curves, the volumes and surfaces of solids of revolution, &c.

Examples.—Areas of Curves, Volumes of Solids, &c.

(1.) The general equation to a parabola of any order is $y^{m+n} = ax m^n$. Then, since $A = \int y\,dx$ taken between the proper limits, we have in this case

$$A = \int a^{\frac{m}{m+n}}\, x^{\frac{n}{m+n}}\, dx$$

$$= \frac{m+n}{m+2n}\, a^{\frac{m}{m+n}}\, x^{\frac{m+2n}{m+n}} + C.$$

Then we perceive that the area $= 0$ when $x = 0$, $\therefore$ $C = 0$; and taking the above between the limits $x = 0$ and $x = x$, since the value of the integral commences when $x = 0$ and ends when $x = x$, we have

$$\int_0^x a^{\frac{m}{m+n}}\, x^{\frac{n}{m+n}}\, dx = \frac{m+n}{m+2n}\, a^{\frac{m}{m+n}}\, x^{\frac{m+2n}{m+n}}.$$

(2.) The general equation to hyperbolas referred to their asymptotes is

$$x^m y^n = a^{m+n},$$

$$\therefore \; y^n = \frac{a^{m+n}}{x^m}, \quad \text{and } \; y = \frac{a^{\frac{m+n}{n}}}{x^{\frac{m}{n}}},$$

$$\therefore \; A = \int y\,dx = \int \frac{a^{\frac{m+n}{n}}}{x^{\frac{m}{n}}}\,dx = \int a^{\frac{m+n}{n}} x^{-\frac{m}{n}}\,dx$$

$$= \frac{n}{n-m}\, a^{\frac{m+n}{n}}\, x^{\frac{n-m}{n}} + C.$$

We must here determine the constant C, as in the above example, that is, find when the value of x makes the area $= 0$, &c. We must here observe, that this formula fails when $m = n$; for then $A = C + a^2 \log x$, which cannot be determined by the above method.

(3.) The equation to the tractrix is

$$\frac{dy}{dx} = -\frac{y}{(a^2 - y^2)^{\frac{1}{2}}},$$

$$\therefore \; y\,dx = -dy(a^2 - y^2)^{\frac{1}{2}},$$

and $A = \int y\,dx = -\int dy\,(a^2 - y^2)^{\frac{1}{2}}$

$$= -\frac{y}{2}(a^2 - y^2)^{\frac{1}{2}} - \frac{a^2}{2}\sin^{-1}\frac{y}{a} + C.$$

Then in order to find the area included by the positive axes, let $y = -a$, observing that $C = 0$,

F

$$\therefore \quad \text{whole area} = \frac{\pi a^2}{4}, \qquad \because \quad \sin^{-1} 1 = \frac{\pi}{2}.$$

Since it is shown in all works on the Differential Calculus that

$$\frac{d\,\text{V}}{dx} = \pi y^2 \text{ and} \frac{d\,\text{S}}{dx} = 2\pi y \sqrt{1 + \left(\frac{dy}{dx}\right)^2},$$

it follows that to find the volumes and surfaces of solids we have to integrate these functions between the proper limits.

Examples.

(1.) To find the volume and surface of a sphere.

The equation to the circle referred to the centre is $y^2 = r^2 - x^2$, where r represents the radius of the sphere; and as one value of r lies wholly above and the other wholly below the axis of x, we must integrate between the limits $x = -r$ and $x = r$; we have

$$\text{V} = \pi \int_{-r}^{+r} (r^2 - x^2)\, dx = \frac{4\pi}{3} r^3.$$

If we integrated this without reference to limits, the expression we should have would give the volume of the segment of a sphere; and we observe that $\text{C} = 0$ when $x = 0$, since the integral becomes 0.

$$\text{Also, S} = 2\pi \int y \sqrt{1 + \left(\frac{dy}{dx}\right)^2} = 2\pi \int y \sqrt{1 + \frac{x^2}{y^2}}$$

$$= 2\pi \int \sqrt{y^2 + x^2}\, dx$$

$$= 2\pi \int_{-r}^{+r} (r^2)^{\frac{1}{2}}\, dx = 2\pi \times 2r^2 = 4\pi r^2.$$

We might form the integral

$$2\pi \int \sqrt{(y^2 + x^2)}\, dx = 2\pi \int \sqrt{r^2}\, dx$$

by writing the value of y^2 given in the equation to the curve.

Hence the integral $= 2\pi r x$, which is the surface of the segment, whose height is x.

(2.) To find the volume and surface of a prolate spheroid formed by the revolution of an ellipse about its major diameter.

The equation to the ellipse is $y^2 = \dfrac{b^2}{a^2} (a^2 - x^2)$, where a and b represent the major and minor semi-axes respectively.

Hence, $V = \pi \displaystyle\int y^2 dx = \pi \dfrac{b^2}{a^2} \displaystyle\int (a^2 - x^2)\, dx$

$$= \pi \frac{b^2}{a^2}\left(a^2 x - \frac{x^3}{3} \right) + C,$$

which is the volume of a spheroidal segment, remarking that $C = 0$. Next we must integrate between the limits $x = -a$ and $x = a$; then we have

$$V = \pi \frac{b^2}{a^2} \int_{-a}^{+a} (a^2 - x^2)\, dx = \frac{4}{3}\pi b^2 a,$$

$$\frac{dS}{dx} = 2\pi \int y \sqrt{1 + \left(\frac{dy}{dx}\right)^2},$$

and $\dfrac{dy}{dx} = -\dfrac{b}{a}\dfrac{x}{(a^2 - x^2)^{\frac{1}{2}}},\quad \therefore 1 + \left(\dfrac{dy}{dx}\right)^2 = 1 + \dfrac{b^2 x^2}{a^2(a^2 - x^2)}.$

If we write e^2 for $\dfrac{a^2 - b^2}{a^2}$ we have

$$1 + \left(\frac{dy}{dx}\right)^2 = \frac{a^2 - e^2 x^2}{a^2 - x^2},$$

$$\therefore \; S = 2\pi \int y \left\{ 1 + \left(\frac{dy}{dx}\right)^2 \right\} dx = \frac{2\pi b e}{a} \int \left(\frac{a^2}{e^2} - x^2\right)^{\frac{1}{2}} dx$$

$$= \frac{\pi b e}{a} \left\{ x \left(\frac{a^2}{e^2} - x^2\right)^{\frac{1}{2}} + \frac{a^2}{e^2} \sin^{-1} \frac{ex}{a} \right\} + C.$$

Here $C = 0$, and if we integrate between the limits $x = -a$ and $x = a$ we have the whole surface

$$= \frac{2\pi a b}{e} \left\{ e(1 - e^2)^{\frac{1}{2}} + \sin^{-1} e \right\}.$$

(3.) To find the volume and surface of an oblate spheroid.
In order to determine the equation to this, we merely have to change a into b, and we have

$$y^2 = \frac{a^2}{b^2}(b^2 - x^2),$$

and $V = \dfrac{\pi a^2}{b^2} \displaystyle\int (b^2 - x^2)\, dx = \dfrac{\pi a^2 x}{b^2} \left(b^2 - \dfrac{x^2}{3}\right) + C.$

Integrating between the limits $x = -b$ and $x = +b$, observing that $C = 0$; for $V = 0$ when $x = 0$, $\therefore$ the whole volume

$$= \frac{\pi a^2}{b^2} \int_{-b}^{+b} (b^2 - x^2)\, dx = \frac{4}{3}\pi a^2 b,$$

and the surface of the oblate sphere may be found in the same manner as the last.

(4.) To find the volume of a circular spindle.

Here let O be the centre of the circle of which ACB is a segment, and let OC = the rad. $= r$ be perpendicular to AB, $OD = a$, $GD = x$ and the chord $AB = 2c$; $GF = y$; (see figure page 103) then we have

$$r^2 = x^2 + (y + a)^2,$$

$$\therefore y^2 = r^2 - a^2 - x^2 - 2ay,$$

$$V = \pi \int y^2 dx = \pi \int (r^2 - a^2 - x^2 - 2ay)\, dx$$

$$= \pi \left\{ (r^2 - a^2) x + \frac{x^3}{3} - 2a \int y dx \right\}$$

$$= \pi \left\{ (r^2 - a^2) x - \frac{x^3}{3} - 2a \times \text{gen. area DGFC} \right\} + C$$

Here $C = 0$, and the above gives the volume of the frustum HECE, the double of which is the whole frustum HFIL. The limits are $x = -c$, and $x = +c$, we have volume of the spindle

$$= 2\pi \left\{ \tfrac{1}{3} c^3 - a \times \text{gen. area ACB} \right\}.$$

(5) To find the volume of an elliptic spindle.

Let ACB be the generating arc of the ellipse, in which $AD = c$, $OD = i$, $MO = a = $ semi-axis major, and $OC = b = $ semi-axis minor; $DG = x$, and $FG = y$. Then from the property of the ellipse c being the

$$a : b :: (a^2 - x^2)^{\frac{1}{2}} \cdot \frac{b}{a} (a^2 - x^2)^{\frac{1}{2}} = PF.$$

Hence $y = \dfrac{b(a^2 - x^2)^{\frac{1}{2}}}{a} - i,$

$$\therefore y^2 = \frac{b^2}{a^2}(a^2 - x^2) - \frac{2bi\sqrt{a^2 - x^2}}{a} + i^2,$$

$$\therefore V = \pi \int y^2 dx$$

$$= \pi \int \left(b^2 - \frac{b^2 x^2}{a^2} - \frac{2bi}{a}\sqrt{a^2 - x^2} + i^2 \right) dx$$

$$= \pi \int \left\{ b^2 - \frac{b^2 x^2}{a^2} - i^2 - 2i \cdot \frac{b}{a}(a^2 - x^2)^{\frac{1}{2}} + 2i^2 \right\} dx$$

$$= \pi \int \left\{ b^2 \cdot \frac{c^2 - x^2}{a^2} - 2iy \right\} dx$$

$$= \pi \left\{ b^2 x \cdot \frac{3c^2 - x^2}{3a^2} - 2i \cdot \text{area DGFC} \right\} + C$$

$$= \text{volume of ECFH, and when } x = c, C = o,$$

$$\therefore \text{ we have } \pi \left\{ b^2 c \cdot \frac{3c^2 - c^2}{3 a^2} - 2i \cdot \text{area DAC} \right\},$$

$$\text{or, } \pi \left\{ \frac{2 b^2}{3 a^2} \cdot c^3 - 2i \cdot \text{area DAC} \right\},$$

the double of which will give the volume of the whole spindle.

(6.) To find the volume of a hyperbolic spindle.

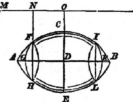

Putting $i =$ the central distance O D, $c =$ A B, and retaining the notation employed in the last, we have by the nature of the curve,

$$\text{OM : OC} = b :: \sqrt{(a^2 + x^2)} : \text{NF} = \frac{b \sqrt{a^2 + x^2}}{a},$$

$$\therefore y = \text{OD} - \text{NF} = i - \frac{b \sqrt{a^2 + x^2}}{a}.$$

$$\therefore V = \int y' \, dx$$

$$= \pi \int \left\{ i^2 - \frac{2 b i}{a} \sqrt{a^2 + x^2} + b^2 \left(\frac{a^2 + x^2}{a^2} \right) \right\} dx$$

$$= \pi \int \left\{ b^2 - i^2 + \frac{b^2 x^2}{a^2} + \left(i - \frac{(a^2 + x^2)^{\frac{1}{2}}}{a} \right) 2i \right\} dx$$

$$= \pi \int \left\{ b^2 - i^2 + \frac{b^2 x^2}{a^2} + 2iy \right\} dx$$

$$= \pi \left\{ 2i \times \text{area GFCD} - \frac{b^2 x}{a^2} \left(\frac{1}{4} c^2 - \frac{x^2}{3} \right) \right\} + C$$

$$= \pi \left\{ 2i \times \text{area GFCD} - \frac{b^2 x}{a^2} \times \left(\frac{3 c^2 - 4 x^2}{12} \right) \right\} + C,$$

and $C = 0$. This is the volume of the frustum FCEH, and for the volume of the spindle we have

$$2\pi \left\{ 2i \times \text{area AFCD} - \frac{b^2 c^3}{12 a^2} \right\}.$$

(7.) To find the surface of a circular spindle.

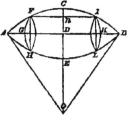

Retaining the notation used in the last, we have

$$no = y + a$$

$$= \sqrt{OF^2 - Fn^2} = \sqrt{r^2 - x^2},$$

$$\therefore y = (r^2 - x^2)^{\frac{1}{2}} - a,$$

and $\dfrac{dy}{dx} = - \dfrac{x}{(r^2 - x^2)^{\frac{1}{2}}}$ and $\left(\dfrac{dy}{dx}\right)^2 = \dfrac{x^2}{r^2 - x^2}$,

$$\therefore S = 2\pi \int y \sqrt{1 + \frac{x^2}{r^2 - x^2}} \, dx$$

$$= 2\pi \int \left\{ (r^2 - x^2)^{\frac{1}{2}} - a \right\} \left\{ \sqrt{\frac{r^2}{r^2 - x^2}} \right\} dx$$

$$= 2\pi r \int \left\{ 1 - \frac{a}{\sqrt{r^2 - x^2}} \right\} dx = 2\pi r \left\{ x - a \sin^{-1} \frac{x}{r} \right\} + C,$$

and $C=0$, the surface of the part HECF; and if $x=c$, we have

$$S = 2\pi r \left\{ c - a \sin^{-1} \frac{c}{r} \right\},$$

the double of which will be the surface of the whole spindle ACBE

(8.) To find the volume of a parabolic spindle.

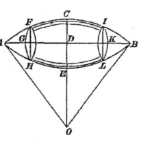

Put $CD = h$, $AB = 2c$, and $x = AG$, and $y = FG$.

From the property of the parabola

$$AD^2 : AG . GB :: CD : EG,$$

$$c^2 : x(2c - x) :: h : y,$$

$$\therefore y = \frac{h(2cx - x^2)}{c^2}, \quad \therefore y^2 = \frac{h^2}{c^4}(4c^2 x^2 - 4cx^3 + x^4),$$

$$\therefore V = \pi \int y^2 dx$$

$$= \pi \frac{h^2}{c^4} \int (4c^2 x^2 - 4cx^3 + x^4)\, dx = \frac{\pi h^2}{c^4}\left(\frac{4c^2 x^3}{3} - cx^4 + \frac{x^5}{5}\right),$$

the volume of AFH, since $C = 0$; when $x = c$, we have

$$\frac{\pi h^2}{c^4}\left(\frac{4c^5}{3} - c^5 + \frac{c^5}{5}\right) = \frac{8}{15}\pi h^2 c =$$

volume of the semi-spindle; and, as we have found the volume of the part AFH, if we subtract this from the whole semi-spindle, we shall have the frustum EHCF, the double of which will be volume of the whole frustum EHIL.

In the same manner as in the last example, we may find the surface of the parabolic spindle.

(9.) In a parabola find the area included between the curve, its evolute, and its radius of curvature.

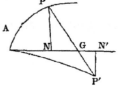

$$\text{area parabola } ANP = \int y\, dx$$

$$= 2\sqrt{a}\int \sqrt{x}\, dx = \frac{4}{3}\sqrt{a}\, x^{\frac{3}{2}},$$

$$\text{area evolute } AN'P' = \int \beta\, d\alpha = \frac{2}{3\sqrt{3a}}\int (\alpha - 2a)^{\frac{3}{2}}\, d\alpha$$

$$= \frac{4}{15\sqrt{3a}}(a-2a)^{\frac{3}{2}} = \frac{4}{15\sqrt{3a}} 9\sqrt{3}x^{\frac{5}{2}} = \frac{12\,x^{\frac{3}{2}}}{5\sqrt{a}}.$$

$$\text{subnormal } NG = y\frac{dy}{dx} = y\frac{2a}{y} = 2a,$$

$$\text{area } PNG = \frac{NP \cdot NG}{2} = ay = 2a^{\frac{3}{2}}x^{\frac{1}{2}}$$

$$GN' = AN' - AG = 3x + 2a - x - 2a = 2x,$$

$$\text{area } P'N'G = \frac{GN' \cdot N'P'}{2} = \beta x = \frac{y^{3}}{4a^{2}}\,x$$

$$= \frac{4ax^{2}}{4a^{2}}2\sqrt{ax} = \frac{2x^{\frac{5}{2}}}{\sqrt{a}},$$

$$\text{area } APP' = APN + NPG + AN'P' - GN'P'$$

$$= \frac{4}{3}\sqrt{ax^{\frac{3}{2}}} + 2a^{\frac{3}{2}}x^{\frac{1}{2}} + \frac{12x^{\frac{3}{2}}}{5\sqrt{a}} - \frac{2x^{\frac{5}{2}}}{\sqrt{a}}$$

$$= \frac{20ax^{\frac{3}{2}} + 30a^{2}x^{\frac{1}{2}} + 6x^{\frac{5}{2}}}{15\sqrt{a}}$$

$$= \frac{2\sqrt{x}}{\sqrt{a}}\left(a^{2} + \frac{2}{3}ax + \frac{1}{5}x^{2}\right).$$

(10.) To find the length of the spiral of Archimedes.

$$r = a \qquad \frac{dr}{d\theta} = a,$$

$$\frac{ds}{dr} = \frac{ds}{d\theta}\cdot\frac{d\theta}{dr} = \frac{1}{a}\sqrt{r^{2} + a^{2}}$$

$$s = \frac{1}{a}\int \sqrt{r^2 + a^2}\, dr = \frac{1}{a}\int \frac{r^2\, dr}{\sqrt{r^2 + a^2}} + a\int \frac{dr}{\sqrt{r^2 + a^2}}$$

$$= \frac{r}{a}\sqrt{r^2 + a^2} - \frac{1}{a}\int dr \sqrt{r^2 + a^2} + a\int \frac{dr}{\sqrt{r^2 + a^2}}$$

$$= \frac{r}{2a}\sqrt{r^2 + a^2} - \frac{a}{2}\int \frac{dr}{\sqrt{r^2 + a^2}} + a\int \frac{dr}{\sqrt{r^2 + a^2}}$$

$$= \frac{r}{2a}\sqrt{r^2 + a^2} + \frac{a}{2}\int \frac{dr}{\sqrt{r^2 + a^2}}$$

$$= \frac{r}{2a}\sqrt{r^2 + a^2} + \frac{a}{2}\log\left(\frac{\sqrt{r^2 + a^2} + r}{a}\right).$$

(11.) On A B, the diameter of a given semicircle A C B, take A D = the chord A C; join C, D; bisect C D in P, and find the equation and area of the locus of P.

Join A, P, and put A P $= y$, $\angle$ PAD $= \varphi$, and AB $= 2\,a$; then A C $=$ A D $= 2\,a \cos 2\varphi$, and A P or $y = 2\,a \cos \varphi \cos 2\varphi$. Let A P $= y$ be drawn in an opposite direction; then φ will become $180° + \phi$, and A P or y will become $= 2\,a \cos(180° = \varphi) \times \cos(360° + 2\phi) = -2\,a \cos \varphi \cos 2\phi = -y$, which indicates that when ϕ passes 180°, the point P will begin at B, and describe exactly the same curve which it has described; hence, when φ arrives at 180°, the curve is complete. When $y = 0$, ϕ has three values from 0° to

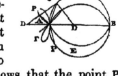

180°, viz., 45°, 90°, and 135°, which shows that the point P returns thrice to the point A, and therefore describes three noduses; 2 area ABP $= \int y^2 d\varphi = 4a^2 \int d\varphi \cos^2 \varphi\, (1 - 4\sin^2 \varphi \cos^2 \varphi) = a^2 \varphi + \frac{1}{3} a^2 \sin \varphi\, (3 \cos \varphi - 2 \cos^3 \varphi + 8 \cos^5 \varphi)$, this between $\varphi = 0°$, and $\varphi = 45°$, gives $\frac{1}{4}$ circle $+ \frac{2}{3} a^2$, or $1\cdot45206\ a^2$ for the area A P B E A, and between $\varphi = 45°$, and $\varphi = 90°$, gives $\frac{1}{4}$ circle $- \frac{2}{3} a^2$, or $\cdot118733'\ a^2$ for the area of the other two small and equal noduses, each of which are therefore $= \frac{1}{8}$ circle $- \frac{1}{3} a^2$, or $\cdot05936'\ a^2$; hence the area of the entire curve comprehending the three noduses is equal to the semicircle A C B.

(12.) A C D B is a given circle, whose diameter is A B; on the chord A D, which is an arithmetical mean between the chord A C and the diameter A B, take A P, a geometrical mean between A C and A B, and show the nature and area of the curve which is the locus of P.

Put A B = a, A P = r, and $\angle$ P A B = θ; then A D = $a \cos \theta$, and A C = 2 A D — A B = $2\,a \cos \theta — a$, and consequently A P^2 or r^2 = A C . A B = $a^2\,(2 \cos \theta — 1)$, the polar equation to the curve. For the quadrature we have

$$2 \text{ area ABP} = \int r^2\,d\theta = a^2 \int d\theta\,(2\cos\theta - 1)$$
$$= 2\,a^2 \sin\theta - a^2\,\theta.$$

To ascertain the limit, put $r = 0$; then r^2 or $a^2\,(2 \cos \theta — 1) = 0$, and $\cos \theta = \frac{1}{2}$; therefore the curve makes an angle of $60°$ with the diameter A B; hence, taking the above between

$\theta = 0$, and $\theta = 60°$, we get $a^2\,\sqrt{3} - \dfrac{1}{6}\,a^2\,.\,2\pi$, or $a^2 \times $ ·6848.

(13.) Let the given circle C E D, whose diameter is C D, touch the indefinite right line A D in D, and from the given point A, draw the right line A E C, on which take A P = the sine of the arc E C, and find the equation and area of the curve which is the locus of P.

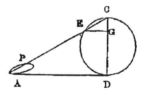

The lines being drawn as enunciated, put C D = a, $\angle$ D A C = ϕ; then A D = $a \cot\phi$, and, by similar $\triangle$s, C D : A D :: C G : E G = $\cot \phi \times$ C G; but, by the circle,

$$E G = (a\,.\,\text{C G} — \text{C G}^2)^{\frac{1}{2}}; \text{ whence C G} = \frac{a}{\cot^2 \phi + 1} = a$$

$\sin^2 \phi$, E G = A P = $(a^2 \sin^2 \phi — a^2 \sin^4 \phi)^{\frac{1}{2}} = \frac{1}{2}\,a \sin 2\phi$; the polar equation of the curve. When $\phi = 90°$, A P = 0, when $\phi = 45°$, A P = $\frac{1}{4}\,a$, and when $\phi = 0°$, A P = 0; hence A is a *punctum duplex*.

Quadrature.— $\int \frac{1}{2} A P \times d\varphi = \frac{1}{8} a^2 \int d\varphi \sin^2 2\varphi = \frac{1}{8} a^2$
$(- \frac{1}{8} \sin 4\varphi + \frac{1}{2}\varphi)$, which, between $\varphi = 90°$, and $\varphi = 0°$, gives $\frac{1}{8}$ of the given circle for the area of the *nodus* A P A

(12.) Let A B, the base of the right-angled triangle A B C, be produced till B P be equal to the diameter of the inscribed circle. Required the equation, quadrature, and greatest possible ordinate of the curve which is the locus of P, the hypothenuse A C being given.

Put the hypothenuse $A C = a$, $3\,14159265 = \pi$, and angle $C A B = x$; then, by trigonometry, A B $= a \cos x$, and $B C = a \sin x$; whence $B P = $ A B + B C − A C $= a (\cos x + \sin x − 1)$, and $A P = a$ $(2 \cos x + \sin x − 1)$, the polar equation of the curve When $x = 90°$, A P = 0, and when $x = 0°$, A P $= a$; ∴ the curve commences at A and terminates at C. The maximum polar ordinate is when $\cos x = 2 \sin x$, or when $\sin x = \frac{1}{5}\sqrt{5}$, or $x = 26° \, 33' \, 9''$.

Quadrature.—The differential of the area is $= \frac{1}{2} A P^2 \, dx$ $= \frac{1}{2} a^2 \, dx \times (4 \cos^2 x + 4 \cos x \sin x + \sin^2 x − 4 \cos x − 2$

$\sin x + 1) = \frac{1}{2} a^2 \, dx \times \left(\frac{7}{2} − 4 \cos x − 2 \sin x + \frac{3}{2} \cos 2x \right.$

$\left. + 2 \sin 2x \right)$, and the integrals give the area $= \frac{1}{2} a^2 \times \left(\frac{7x}{2} \right.$

$\left. − 4 \sin x + 2 \cos x + \frac{3}{4} \sin 2x − \cos 2x \right)$; which, be-

tween $x = 0°$, and $x = 90°$, gives $\frac{7}{8} a^2 \pi − 2 a^2$ for the area of the whole curve A P C, or in numbers $= \cdot 74889357 \, a^2$, as required.

The area of the space inclosed by the curve

$$a y^2 = x^0 \sqrt{a^2 − x^2} \text{ is } \frac{8}{5} a^2.$$

$$\text{Area} = \int y \, dx = \frac{1}{\sqrt{a}} \int x \, (a^2 − x^2)^{\frac{1}{2}} \, dx$$

$$= − \frac{1}{\sqrt{a}} \cdot \frac{2}{5} (a^2 − x^2)^{\frac{3}{4}} + C$$

If $x = 0$ $\qquad$ Area $= \dfrac{2}{5} a^2 + $ C.

If $x = a$ $\qquad$ Area $= 0,$

$\qquad$ $\therefore$ Whole area $= \dfrac{4 \cdot 2}{5} a^2.$

The length of the curve, $y \log \left(\dfrac{e^x + 1}{e^x - 1} \right)$

$\qquad$ from $x = 1$ to $x = 2$ is $= \log (e + e^{-1}).$

$$y = \log \left(\dfrac{e^x + 1}{e^x - 1} \right),$$

$$\dfrac{dy}{dx} = \dfrac{e^x (e^x - 1) - e^x (e^x + 1)}{(e^x - 1)^2} \cdot \dfrac{e^x - 1}{e^x + 1}$$

$$= \dfrac{-2 e^x}{e^{2x} - 1},$$

$$1 + \dfrac{dy^2}{dx^2} = \dfrac{e^{4x} - 2 e^{2x} + 1 + 4 e^{2x}}{(e^{2x} - 1)^2} = \left(\dfrac{e^{2x} + 1}{e^{2x} - 1} \right)^2,$$

$$\therefore \ \text{S} = \int dx \sqrt{1 + \dfrac{dy^2}{dx^2}} = \int \dfrac{e^{2x} + 1}{e^{2x} - 1} dx$$

$$= \int \dfrac{e^x \, dx}{e^x + 1} + \int \dfrac{d x}{e^x - 1} = \int \dfrac{e^x \, dx}{e^x + 1} + \int \dfrac{e^x \, dx}{e^x - 1} - \int d x$$

$$\text{S} = \log (e^{2x} - 1) - x + \text{C}.$$

If $x = 2$ $\qquad$ S $= \log (e^4 - 1) - 2 + $ C.

$\qquad x = 1$ $\qquad$ S $= \log (e^2 - 1) - 1 + $ C,

$\therefore$ from $x = 2$ to $x = 1$ $s = \log (e^4 - 1) - \log (e^2 - 1) - 1,$

$$s = \log (e^2 + 1) - \log e$$

$$= \log \left(\frac{e^2 + 1}{e} \right) = \log (e + e^{-1}).$$

(14.) The length of the curve $8\,a^3 y = x^4 + 6\,a^2 x^2$, measured from the origin of co-ordinates, is

$$\frac{x}{8\,a^3} (x^2 + 4\,a^2)^{\frac{3}{2}}$$

$$y = \frac{x}{8\,a^3} (x^4 + 6\,a^2 x^2)$$

$$\frac{dy}{dx} = \frac{1}{8\,a^3} (4x^3 + 12\,a^2 x) = \frac{1}{2\,a^3} (x^2 + 3\,a^2 x)$$

$$1 + \frac{dy^2}{dx^2} = \frac{1}{4\,a^6} (x^6 + 6\,a^2 x^4 + 9\,a^4 x^2 + 4\,a^6)$$

$$= \frac{1}{4\,a^6} \{ x^6 + 3\,a^2 x^4 + 3\,a^4 x^2 + a^6 + 3\,a^2 (x^4 + 2\,a^2 x^2 + a^4) \}$$

$$= \frac{1}{4\,a^6} \left\{ \left(x^2 + a^2 \right)^3 + 3\,a^2 (x^2 + a^2)^2 \right\}$$

$$= \frac{(x^2 + a^2)^2}{4\,a^6} (x^2 + 4\,a^2),$$

$$\therefore s = \int dx \sqrt{1 + \frac{dy^2}{dx^2}} = \frac{1}{2\,a^3} \int (x^2 + a^2) \sqrt{x^2 + 4\,a^2}\, dx$$

$$= \frac{1}{2\,a^3} \frac{x}{4} (x^2 + 4\,a^2)^{\frac{3}{2}}.$$

(15.) The volume generated by the curve $y^2 (x - 4a) = ax (x - 3a)$ revolving about the axis of x, from $x = 0$ to $x = 3a$ is $= \frac{1}{2} \pi a^3 (15 - 16 \log 2)$.

$$y^2 = a \frac{x^2 - 3ax}{x - 4a} = a \left(x + a + \frac{4a^2}{x - 4a} \right),$$

$$\therefore V = \pi \int y^2 \, dx = \pi a \left(\frac{x^2}{2} + ax + 4a^2 \log (x - 4a) \right) + C.$$

If $x = 0$ \qquad $0 = 4\pi a^3 \log (-4a) + C.$

If $x = 3a$

$$0 = \pi a \left(\frac{9a^2}{2} + 3a^2 + 4a^2 \log (-a) \right) + C,$$

$$\therefore V = \pi a \left\{ \frac{15a^2}{2} - 4a^2 \left\{ \log (-4a) - \log (-a) \right\} \right\}$$

$$= \frac{\pi}{2} a^3 \left\{ 15 - 8 \log 4 \right\}$$

$$= \frac{\pi}{2} a^3 \left\{ 15 - 16 \log 2 \right\}.$$

(16.) Find the area of the curve in which

$$r = \frac{(a^2 - b^2) \sin \theta \cos \theta}{\sqrt{a^2 \sin^2 \theta + b^2 \cos^2 \theta}}$$

$$\text{area} = \frac{1}{2} \int r^2 \, d\theta = \frac{1}{2} \int \frac{(a^2 - b^2)^2 \sin^2 \theta \cos^2 \theta}{a^2 \sin^2 \theta + b^2 \cos^2 \theta} \, d\theta$$

$$= \frac{1}{2} \int \frac{(a^2 - b^2)^2 \sin^2 \theta \, d\theta}{a^2 \tan^2 \theta + b^2}.$$

Let $x = \tan\theta$, $d\theta = \dfrac{1}{1+x^2}\,dx$ $\sin\theta = \dfrac{x}{\sqrt{1+x^2}}$

$$\text{area} = \frac{1}{2}\int \frac{(a^2-b^2)^2\,\dfrac{x^2}{1+x^2}\,\dfrac{1}{1+x^2}\,dx}{a^2x^2+b^2}$$

$$= \frac{1}{2}\int \frac{(a^2-b^2)^2\,x^2\,dx}{(a^2x^2+b^2)(1+x^2)^2}$$

Let $\dfrac{U}{V} = \dfrac{A}{a^2x^2+b^2} + \dfrac{B}{(1+x^2)^2} + \dfrac{C}{1+x^2}$

$$(a^2-b^2)^2\,x^2$$

$$= A(1+x^2)^2 + B(a^2x^2+b^2) + C(1+x^2)(a^2x^2+b^2).$$

Let $x^2 = -\dfrac{b^2}{a^2}$, $\therefore -(a^2-b^2)^2\dfrac{b^2}{a^2} = A\left(\dfrac{a^2-b^2}{a^2}\right)^2$

$$A = -a^2b^2.$$

Let $x = \sqrt{-1}$; $-(a^2-b^2)^2 = -B(a^2-b^2)$,

$$\therefore B = a^2 - b^2,$$

$$\therefore (a^2-b^2)^2\,x^2 - (a^2-b^2)(a^2x^2+b^2) + a^2b^2(1+x^2)^2$$

$$= C(1+x^2)(a^2x^2+b^2)$$

$$-(a^2-b^2)(1+x^2)b^2 + a^2b^2(1+x^2)^2 = C(1+x^2)$$
$$(a^2x^2+b^2)$$

$$b^2(1+x^2)(a^2+a^2x^2-a^2+b^2) = C(1+x^2)(a^2x^2+b^2)$$

$$C = b^2,$$

$$\therefore \int \frac{(a^2 - b^2)\, x^2\, d x}{(a^2 x^2 + b^2)\,(1 + x^2)^2}$$

$$= -\, a^2 b^2 \int \frac{d x}{a^2 x^2 + b^2} + (a^2 - b^2) \int \frac{d x}{(1 + x^2)^2}$$

$$+\, b^2 \int \frac{d x}{1 + x^2}.$$

But $\displaystyle \int \frac{d x}{(1 + x^2)^2} = \int \frac{d x}{1 + x^2} - \int \frac{x^2\, d x}{(1 + x^2)^2}$

$$= \int \frac{d x}{1 + x^2} + \frac{x}{2\,(1 + x^2)} - \frac{1}{2} \int \frac{d x}{1 + x^2}$$

$$= \frac{1}{2} \int \frac{d x}{1 + x^2} + \frac{x}{2\,(1 + x^2)},$$

$$\therefore \; \text{area} = -\frac{a^2 b^2}{2} \int \frac{d x}{a^2 x^2 + b^2} + \frac{a^2 + b^2}{4} \int \frac{d x}{1 + x^2}$$

$$+\, \frac{x}{4\,(1 + x^2)}$$

$$= -\frac{d b}{2} \tan^{-1} \frac{a x}{b} + \frac{a^2 + b^2}{4} \tan^{-1} x + \frac{x}{4\,(1 + x^2)}.$$

Taking this between limits $\theta = 0$ and $\theta = 90$;

or, $x = 0 \quad x = \infty$

$$\text{area} = \frac{a^2 + b^2}{4} \frac{\pi}{2} - \frac{a b}{2} \frac{\pi}{2} = \frac{(a - b)^2}{4} \frac{\pi}{2}.$$

(17.) The length of the epicycloid after one revolution of the generating circle $= 8\dfrac{b}{a}(a+b)$, and the area between the epicycloid and the circle $= \pi b^2\left(3+\dfrac{2b}{a}\right)$

$p^2 = c^2\left(\dfrac{r^2-a^2}{c^2-a^2}\right)$, equation to epicycloid in terms of rad. vector and perp. on tangent where $c = a+2b$.

$$\frac{ds}{dr} = \frac{r}{\sqrt{r^2-p^2}}$$

$$r^2-p^2 = \frac{c^2r^2-a^2r^2-c^2r^2+a^2c^2}{c^2-a^2} = \frac{a^2(c^2-r^2)}{c^2-a^2}$$

$$\frac{ds}{dr} = \frac{\sqrt{c^2-a^2}}{a}\;\frac{r}{\sqrt{c^2-r^2}}$$

$$s = \pm\frac{\sqrt{c^2-a^2}}{a}\sqrt{c^2-r^2}+\text{C}.$$

If $r = a+2b = c$, then $s = 0$, $\text{C} = 0$.

If $r = a$, then $s = \dfrac{c^2-a^2}{a} = \dfrac{a^2+4ab+4b^2-a^2}{a}$

$$s = \pm\frac{4b}{a}(a+b),$$

hence whole length of arc of epicycloid

$$= \frac{8b}{a}(a+b).$$

Also to find area $\dfrac{d\theta}{dr} = \dfrac{p}{r\sqrt{r^2 - p^2}}$

$$r^2 d\theta = \frac{pr\,dr}{\sqrt{r^2 - p^2}} = \frac{cr\sqrt{r^2 - a^2}}{\sqrt{c^2 - a^2}} \times \frac{\sqrt{c^2 - a^2}\,dr}{a\sqrt{c^2 - r^2}}$$

$$= \frac{c}{a}\frac{r\sqrt{r^2 - a^2}\,dr}{\sqrt{c^2 - r^2}}$$

$$\text{area} = \frac{1}{2}\int r^2\,d\theta = \frac{c}{2a}\int\frac{r\,dr\,\sqrt{r^2 - a^2}}{\sqrt{c^2 - r^2}}.$$

Let $r^2 - a^2 = z^2$, $r\,dr = z\,dz$, $c^2 - r^2 = c^2 - a^2 - z^2$,

$$\text{area} = \frac{c}{2a}\int\frac{z^2\,dz}{\sqrt{\beta^2 - z^2}};\quad \text{if } c^2 - a^2 = \beta^2$$

$$= -\frac{c}{2a}z\sqrt{\beta^2 - z^2} + \frac{c}{2a}\int dz\sqrt{\beta^2 - z^2}$$

$$= -\frac{c}{2a}z\sqrt{\beta^2 - z^2} + \frac{c\beta^2}{2a}\int\frac{dz}{\sqrt{\beta^2 - z^2}}$$

$$-\frac{c}{2a}\int\frac{z^2\,dz}{\sqrt{\beta^2 - z^2}},$$

$$\text{area} = -\frac{c}{4a}z\sqrt{\beta^2 - z^2} + \frac{c\beta^2}{4a}\sin^{-1}\frac{z}{\beta} + C.$$

If $r = a$ then $z = 0$, and area $= 0$,

$r = c$ then $\beta^2 - z^2 = 0$ $z^2 = c^2 - a^2$,

$$\therefore \text{ semi-area} = \frac{c\,(c^2 - a^2)}{4\,a}\, \sin^{-1} \frac{\sqrt{c^2 - a^2}}{\sqrt{c^2 - a^2}}$$

$$= \frac{c\,(c^2 - a^2)}{4\,a}\, \frac{\pi}{2}$$

$$\text{area circle} = \frac{a}{2}\, 2\pi b = \pi a b,$$

$$\text{area between epicycloid and circle} = \frac{c\,(c^2 - a^2)}{4\,a}\, \pi - \pi a b$$

$$= \frac{(a + 2b)\,4\,b\,(a + b)}{4\,a}\, \pi - \pi a b$$

$$= \frac{\pi}{a}\,(a^2 b + 3\,ab + 2\,b^3 - a^2 b)$$

$$= \pi b^2 \left(3 + \frac{2b}{a} \right).$$

(18.) Find the length of the curve where

$$x^{\frac{2}{3}} + y^{\frac{2}{3}} = a^{\frac{2}{3}},$$

$$s = \int \sqrt{1 + \frac{dy^2}{d\,x^2}} = \int \sqrt{1 + \frac{y^{\frac{2}{3}}}{x^{\frac{2}{3}}}}$$

$$= \int \sqrt{\frac{x^{\frac{2}{3}} + y^{\frac{2}{3}}}{x^{\frac{2}{3}}}} = \int \frac{a^{\frac{1}{3}}}{x^{\frac{1}{3}}} = \frac{3}{2}\, a^{\frac{1}{3}}\, x^{\frac{2}{3}}$$

Taking it between the limits $x = 0$, $x = a$,

$$s = \frac{3}{2}\, a.$$

The whole length of the curve $4 \times \frac{3}{2}\, a = 6\,a$.

(19.) If $h =$ height of a parabolic frustum, a and b the radii of the ends, show that

$$\text{Frustum} = \frac{\pi h}{2}(a^2 + b^2).$$

Equation to parabola $y^2 = 4mx$.

$$V = \pi \int_x^{x+h} y^2\, dx = 4\pi m \int_x^{x+h} x\, dx$$

$$= 2\pi m\, \{(x+h)^2 - x^2\} = 2\pi m\,(2xh + h^2)$$

$$\bullet = \frac{\pi h}{2}\{4mx + 4m\,(x+h)\} = \frac{\pi h}{2}(a^2 + b^2).$$

(20.) Find the area of the catenary

$$y = \frac{a}{2}\left(e^{\frac{x}{a}} + e^{-\frac{x}{a}}\right).$$

$$\text{Area} = \int y\, dx = \frac{a}{2}\int \left(e^{\frac{x}{a}} + e^{-\frac{x}{a}}\right) dx$$

$$= \frac{a}{2}\left(a\, e^{\frac{x}{a}} - a\, e^{-\frac{x}{a}}\right),$$

$$= \frac{a}{2}\sqrt{a^2 e^{\frac{2x}{a}} + 2a^2 + a^2 e^{-\frac{2x}{a}} - 4a^2}$$

$$= \frac{a}{2}\sqrt{4y^2 - 4a^2} = a\sqrt{y^2 - a^2}$$

(21.) Find the area of $x^4 y^4 - a^4 y^4 = a^8$,

$$y = \frac{a^2}{\sqrt[4]{x^4 - a^4}}.$$

$$\text{area} = \int y\, dx = a^2 \int \frac{dx}{\sqrt[4]{x^4 - a^4}} = a^2 \int \frac{dx}{x\, z},$$

If $x^4 - a^4 = x^4 z^4$,

$$x = \frac{a}{\sqrt[4]{1 - z^4}},$$

$$\log x = \log a - \frac{1}{4} \log (1 - z^4),$$

$$\frac{dx}{x} = \frac{z^3}{1 - z^4}\, dz,$$

$$\therefore \text{area} = a^2 \int \frac{1}{z}\, \frac{z^3}{1 - z^4}\, dz = a^2 \int \frac{z^2}{1 - z^4}\, dz$$

$$= \frac{a^2}{2} \left\{ \int \frac{1}{1 - z^2} - \int \frac{1}{1 + z^2} \right\}$$

$$= \frac{a^2}{4} \int \frac{1}{1 + z} + \frac{a^2}{4} \int \frac{1}{1 - z} - \frac{a^2}{2} \int \frac{1}{1 + z^2}$$

$$= \frac{a^2}{4} \log \frac{1 + z}{1 - z} - \frac{a^2}{2} \tan^{-1} z.$$

But $z^4 = \dfrac{x^4 - a^4}{x^4} = \dfrac{a^8}{x^4 y^4}$,

$$\therefore \text{area} = \frac{a^2}{4} \log \frac{1 + \dfrac{a^2}{xy}}{1 - \dfrac{a^2}{xy}} - \frac{a^2}{2} \tan^{-1} \frac{a^2}{xy}$$

$$= \frac{a^2}{2} \left\{ \log \sqrt{\frac{xy + a^2}{xy - a^2}} - \tan^{-1} \frac{a^2}{xy} \right\}.$$

(22.) Find the volume generated by the revolution of the witch round its asymptote.

$$y^2 = 4a^2 \frac{2a - x}{x} \quad \text{equation to witch,}$$

$$xy^2 = 8a^3 - 4a^2x,$$

$$x = \frac{8a^3}{y^2 + 4a^2},$$

$$\text{Volume} = \pi \int x^2 \, dy = 16a^4 \, \pi \int \frac{4a^2 \, dy}{(y^2 + 4a^2)^2}$$

$$= 16a^4 \pi \int \frac{dy}{y^2 + 4a^2} - 16a^4 \, \pi \int \frac{y^2 \, dy}{(y^2 + 4a^2)^2}.$$

$$\text{But} \int y \frac{y \, dy}{(y^2 + 4a^2)^2} = -\frac{1}{2} \frac{y}{y^2 + 4a^2} + \frac{1}{2} \int \frac{dy}{y^2 + 4a^2}.$$

$$\text{Volume} = 8a^4 \pi \frac{y}{y^2 + 4a^2} + 8a^4 \pi \int \frac{dy}{y^2 + 4a^2}$$

$$= 8a^4 \pi \frac{y}{y^2 + 4a^2} + 4a^3 \pi \tan^{-1} \frac{y}{2a}.$$

Taking between the limits of $y = \infty$ and $y = 0$.

Volume $= 4a^3 \pi \tan^{-1} \infty = 2a^3 \pi^2$, and whole volume generated by curve both above and below the abscissa $= 4a^3 \pi^2$.

MISCELLANEOUS EXAMPLES.

(1.) $\int \sqrt{2ax - x^2} \cdot dx$

$$= \int \frac{(2ax - x^2)\, dx}{\sqrt{2ax - x^2}}$$

$$= \int \frac{x\, dx\, \{a + (a - x)\}}{\sqrt{2ax - x^2}}$$

$$= \int \frac{ax\, dx}{\sqrt{2ax - x^2}} + \int \frac{(a - x)\, x\, dx}{\sqrt{2ax - x^2}}.$$

$$\int \frac{(a - x)\, dx \cdot x}{\sqrt{2ax - x^2}}$$

$$= x \sqrt{2ax - x^2} - \int \sqrt{2ax - x^2}\, dx,$$

$$\int \frac{ax\, dx}{\sqrt{2ax - x^2}} = -\int \frac{\{(a - x)\, a - a^2\}\, dx}{\sqrt{2ax - x^2}}$$

$$= -a \sqrt{2ax - x^2} + a^2 \cdot \text{ver sin}^{-1} \frac{x}{a},$$

$$\therefore \; 2 \int \sqrt{2ax - x^2} \cdot dx$$

$$= (x - a) \sqrt{2ax - x^2} + a^2 \, \text{ver sin}^{-1} \frac{x}{a},$$

$$\therefore \int \sqrt{2ax - x^2} \cdot dx$$

$$= \frac{(x-a)\sqrt{2ax-x^2}}{2} + \frac{a^2}{2} \text{ ver sin}^{-1} \frac{x}{a}.$$

This integral is used by Earnshaw, in finding the centre of gravity of the area of a cycloid.

$$(2) \qquad \int \frac{x^2\, dx}{\sqrt{2ax + x^2}}.$$

By formula of Reductioh,

$$\int \frac{x^n\, dx}{\sqrt{2ax + x^2}}$$

$$= \frac{x^{n-1}\sqrt{2ax + x^2}}{n} - \frac{a(2n - 1)}{n} \int \frac{x^{n-1}\, dx}{\sqrt{2a + x^2}}.$$

If $n = 2$, this gives $\displaystyle \int \frac{x^2\, dx}{\sqrt{2ax + x^2}}$

$$= \frac{x\sqrt{2ax + x^2}}{2} - \frac{3a}{2} \int \frac{x\, dx}{\sqrt{2ax + x^2}}.$$

Also $\displaystyle \int \frac{x\, dx}{\sqrt{2ax + x^2}}$

$$= \sqrt{2ax + x^2} - a \int \frac{dx}{\sqrt{2ax + x^2}}.$$

G

And $\int \dfrac{d\,x}{\sqrt{2\,a\,x+x^2}} = \log\{x + a + \sqrt{2\,a\,x + x^2}\},$

$$\therefore \int \dfrac{x\,d\,x}{\sqrt{2\,a\,x + x^2}}$$

$$= \sqrt{2\,a\,x + x^2} - a \log\{a + a + \sqrt{2\,a\,x + x^2}\}$$

$$\int \dfrac{x^2\,d\,x}{\sqrt{2\,a\,x + x^2}} = \dfrac{x\sqrt{2\,a\,x + x^2}}{2} - \dfrac{3\,a}{2}\sqrt{2\,a\,x + x^2}$$

$$+ \dfrac{3\,a^2}{2} \log\{x + a + \sqrt{2\,a\,x + x^2}\}.$$

(3.) $$\int_a^0 \dfrac{x^2\,d\,x}{(2\,a\,x - x^2)^{\frac{3}{2}}}$$

let $p = x \quad dp = dx, \quad dq = \dfrac{x\,d\,x}{(2\,a\,x - x^2)^{\frac{3}{2}}},$

$$\therefore q = \int \dfrac{x\,d\,x}{(2\,a\,x - x^2)^{\frac{3}{2}}} = \int (2\,a - x)^{-\frac{3}{2}}\,x^{\frac{1}{2}}\,d\,x =$$

$$\int (2\,a\,x^{-1} - 1)^{-\frac{3}{2}}\,x^{-2}\,dx$$

$$= -\dfrac{1}{2\,a}\int (2\,a\,x^{-1} - 1)^{-\frac{3}{2}} \times - 2\,a\,x^{-2}\,dx$$

$$= \dfrac{(2\,a\,x^{-1} - 1)^{-\frac{1}{2}}}{a} = \dfrac{1}{a}\sqrt{\dfrac{x}{2\,a - x}}.$$

$$\int p\,dq = pq - \int q\,dp$$

$$= \frac{x}{a}\sqrt{\frac{x}{2a-x}} - \frac{1}{a}\int \frac{\sqrt{x}\,.\,dx}{\sqrt{2a-x}}$$

$$= \frac{x}{a}\sqrt{\frac{x}{2a-x}} - \frac{1}{a}\int \frac{x\,dx}{\sqrt{2ax-x^2}};$$

but by the formula of reduction,

$$\int \frac{x\,dx}{\sqrt{2ax-x^2}} = -\sqrt{2ax-x^2} + a\,\mathrm{vers}^{-1}\frac{x}{a},$$

$$\therefore \int \frac{x^2\,dx}{(2ax-x^2)^{\frac{3}{2}}}$$

$$= \frac{x}{a}\sqrt{\frac{x}{2a-x}} + \frac{1}{a}\sqrt{2ax-x^2} - \mathrm{vers}^{-1}\frac{x}{a},$$

taken between the limits of 0 and a,

$$\int_a^0 \frac{x^2\,dx}{(2ax-x^2)^{\frac{3}{2}}} = 1 + 1 - \frac{\pi}{2} = 2 - \frac{\pi}{2} = \cdot4292.$$

This integral is used in Barlow on the strength of mate‑ rials. See pages 364, 365.

$$(4.)\ \int \frac{x^2\,dx}{x^4+a^4} = \left(\frac{1}{4a}\log\frac{x-a}{x+a} + \frac{1}{2a}\tan^{-1}\frac{x}{a}\right)\frac{1}{a}.$$

$$(5.)\ \int \frac{dx}{(x^2+b)^2} = \frac{x}{2b(x^2+b)} + \frac{1}{2b\sqrt{b}}\tan^{-1}\frac{x}{\sqrt{b}}.$$

(6) $$\int \frac{d\,x}{(x+a)\,(x^2+b)^2}$$

$$= \frac{1}{(a^2+b)^2}\,\log\frac{x+a}{\sqrt{x^2+b}} + \frac{1}{2b\,(a^2+b)}\cdot\frac{ax+b}{x^2+b}$$

$$+ \frac{a\,(a^2+3b)}{2\,b\,\sqrt{b}\,(a^2+b)^2}\cdot\tan^{-1}\frac{x}{\sqrt{b}},$$

(7.) $$\int \frac{x^4\,d\,x}{(a+bx^2)^3}$$

$$= -\left(\frac{5x^3}{8b}+\frac{3\,a\,x}{8b^2}\right)\cdot\frac{1}{(a+b\,x)^2} + \frac{3}{8b^2}\int \frac{d\,x}{a+b\,x^2}$$

(8.) $$\int \frac{x^2\,d\,x}{(a+bx)^2} = \frac{1}{b^3}(a+bx) - \frac{a^2}{b^3}\cdot\frac{1}{a+bx} - \frac{2a}{b^3}\log\,(a+b\,x)$$

The formula of reduction

(9.) $$\int \frac{d\,x}{x^m\,\sqrt{1-x^2}} = -\frac{\sqrt{1-x^2}}{(m-1)\,x^{m-1}}$$

$$+ \frac{m-2}{m-1}\int \frac{d\,x}{x^{m-2}\,\sqrt{1-x^2}},$$

gives the following integrals, taking m odd

$$\int \frac{d\,x}{x\,\sqrt{1-x^2}} = -\log\left(\frac{1+\sqrt{1-x^2}}{x}\right) + C$$

$$\int \frac{d\,x}{x^3\,\sqrt{1-x^2}} = -\frac{\sqrt{1-x^2}}{2\,x^2} + \frac{1}{2}\int \frac{d\,x}{x\,\sqrt{1-x^2}}$$

$$\int \frac{d\,x}{x^5\,\sqrt{1-x^2}} = -\frac{\sqrt{1-x^2}}{4\,x^4} + \frac{3}{4}\int \frac{d\,x}{x^3\,\sqrt{1-x^2}}$$

$$\int \frac{dx}{x^7 \sqrt{1-x^2}} = -\frac{\sqrt{1-x^2}}{6x^6} + \frac{5}{6} \int \frac{dx}{x^5 \sqrt{1-x^2}}$$

Taking m even

$$\int \frac{dx}{x^2 \sqrt{1-x^2}} = -\frac{\sqrt{1-x^2}}{x} + C$$

$$\int \frac{dx}{x^4 \sqrt{1-x^2}} = -\frac{\sqrt{1-x^2}}{3x^3} + \frac{2}{3} \int \frac{dx}{x^2 \sqrt{1-x^2}},$$

&c.

The following integrals may be done by the general form

$$\int x^{m-1}(a+bx^n)^{\frac{p}{q}}.$$

(10.) $\quad \displaystyle\int \frac{dx}{x \sqrt{a+bx}} = \frac{1}{\sqrt{a}} \log \frac{a+bx-\sqrt{a}}{a+bx+\sqrt{a}}$

(11.) $\quad \displaystyle\int \frac{dx}{x \sqrt{4+3x}} = \frac{\sqrt{4+3x}}{4x} - \frac{3}{16} \log \frac{\sqrt{4+3x}-2}{\sqrt{4+3x}+2}.$

(12.) $\quad \displaystyle\int \frac{dx}{x \sqrt{bx-a}} = \frac{2}{\sqrt{a}} \cdot \sin^{-1} \sqrt{\frac{bx-a}{bx}}.$

(13.) $\quad \displaystyle\int \frac{dx}{x(a+bx)^{\frac{3}{2}}} = \frac{2}{a \sqrt{a+bx}}$

$$+ \frac{1}{a \sqrt{a}} \log \frac{\sqrt{a+bx}-\sqrt{a}}{\sqrt{a+bx}+\sqrt{a}}.$$

(14.) $\quad \displaystyle\int \frac{x^2 dx}{\sqrt[3]{2+3x}} = (2+3x)^{\frac{2}{3}} \left\{ \frac{4-4x+5x^2}{40} \right\}.$

(15.) $\quad \displaystyle\int \frac{dx}{(a+bx^2)^{\frac{5}{2}}} = \left(\frac{1}{3a(a+bx^2)} + \frac{2}{3a^2} \right) \frac{x}{\sqrt{a+bx^2}}.$

(16.) $\quad \displaystyle\int \frac{x^{\frac{1}{2}} dx}{(1+x^{\frac{1}{2}})^{\frac{1}{2}}} = 4(1+x^{\frac{1}{2}})^{\frac{1}{2}} \left\{ \frac{1}{5}(1+x^{\frac{1}{2}})^2 - \frac{2}{3}(1+x^{\frac{1}{2}}) + 1 \right\}.$

(17.) $\displaystyle\int\frac{x^{3n-1}\,dx}{(a+bx^n)^{\frac{1}{4}}} = \frac{2\,(a+bx^n)^{\frac{1}{2}}}{15\,n\,b^3}\,(3\,b^2x^{2n} - 4\,abx^n + 8\,a^2)$

(18.) $\displaystyle\int\frac{dx}{(2\,ax+x^2)^{\frac{5}{2}}} = -\left\{\frac{2}{3\,(2\,ax+x^2)} - \frac{2}{3\,a^2}\right\}$

$$\frac{x+a}{a^2\sqrt{2\,ax+x^2}}.$$

(19.) $\displaystyle\int\frac{e^x\,x\,dx}{(1+x)^2} = \frac{e^x}{1+x}.$

(20.) Show that $e^{\int\frac{d\theta}{\sin\theta}} = \tan\dfrac{\theta}{2}$

(21.) $\displaystyle\int e^{\sqrt{x}}\,x\,dx = 2\,e^{\sqrt{x}}\,(x^{\frac{3}{2}} - 3\,x + 6\sqrt{x} - 6).$

(22.) $\displaystyle\int e^{ax}\sin^2 x\,dx = \frac{2\,e^{ax}}{a\,(a^2+4)}$

$$\dotplus\ \frac{e^{ax}.\sin x}{a^2+4}\,(a\sin x + 2\cos x),$$

(23.) $\displaystyle\int e^x\cos^2 x = \frac{e^x}{2} + \frac{e^x\cos 2x}{10} + \frac{e^x\sin 2x}{5}.$

(24.) $\displaystyle\int\frac{dx}{\sin^2 x\cos^4 x} = 2\tan x + \frac{\tan^3 x}{3} + \frac{1}{\tan x};$

also, show that the integral is

$$\frac{1}{3}\cdot\frac{1}{\sin x\cos^3 x} - \frac{8}{3}\cot 2x*.$$

(25.) $\displaystyle\int\frac{d\theta}{1-\epsilon^2\cos^2\theta} = \frac{1}{\sqrt{1-\epsilon^2}}\tan^{-1}\left\{\frac{\tan\theta}{\sqrt{1-\epsilon^2}}\right\},\ (\epsilon<1).$

* It is observed by De Morgan, that we are liable, in using artifices of integration to produce results which appear different, but which in fact only differ by a constant. This discrepancy does not appear when the integrals are taken between definite limits, since $\varphi a - \varphi b = \varphi a + C - (\varphi b + C)$, are the same. See De Morgan's Differential and Integral Calculus, p. 116.

(26.) $\displaystyle\int \frac{d\,\theta}{1 - \iota^2 \cos^2 \theta} = \frac{1}{2\sqrt{1-\iota^2}} \log \left\{ \frac{\sqrt{\iota^2-1}\,\cos\theta - \sin\theta}{\sqrt{\iota^2-1}\,\cos\theta + \sin\theta} \right\}$

$$(\iota > 1).$$

(27.) $\displaystyle\int \frac{d\,\theta}{\sin^3 (a\,\theta + b)\, \cos^2 (a\,\theta + b)} = \frac{3}{2\,a} \log \tan \left\{ \frac{a\,\theta + b}{2} \right\}$

$$+ \frac{1}{a} \left\{ \frac{1}{\cos (a\,\theta + b)} - \frac{\cos a\,\theta + b}{2\,(\sin^2 a\,\theta + b)} \right\}.$$

(28.) $\displaystyle\int \frac{d\,\theta}{\sec \theta \cos \sec \theta} = \frac{\text{vers } 2\,\theta}{4},$

(29.) $\displaystyle\int d\,\theta \cos\sec 2\,\theta = \frac{1}{2} \log \tan \theta,$

(30.) $\displaystyle\int \frac{d\,\theta}{\cos^2 n\,\theta} = \frac{1}{n} \tan n\,\theta,$

(31.) $\displaystyle\int \frac{d\,\theta}{1 - \tan^2 \theta} = \theta + \frac{1}{8} \log \frac{1 + \sin 2\,\theta}{1 - \sin 2\,\theta}$

(32.) $\displaystyle\int \frac{d\,\theta}{\sin^4 \theta \cos \theta} = \frac{1}{3 \sin^3 \theta} - \frac{1}{\sin \theta} + \frac{1}{2} \log \tan \left(45 + \frac{\theta}{2} \right)$

(33.) $\displaystyle\int \frac{\sin^3 \theta\, d\,\theta}{\cos^2 \theta} = \cos \theta + \sec \theta,$

(34.) $\displaystyle\int \frac{d\,\theta}{\sin \theta \cos^3 \theta} = \frac{1}{2 \cos^2 \theta} + \log \tan \theta,$

(35.) $\displaystyle\int \frac{\sin^3 \theta\, d\,\theta}{\cos^8 \theta} = \frac{\sin^2 \theta}{5 \cos^7 \theta} - \frac{2}{5 \cdot 7 \cos^7 \theta}.$

(36.) $\displaystyle\int_r^R \int_\varphi^{\theta'} r^2 \sin \theta\, d\,\theta\, d\,r = -\frac{1}{3}(R^3 - r^3)\,(\cos \theta - \cos \theta).$

This integral is used by Professor Moseley on the arch. See the "Mechanical Principles of Engineering and Architecture," page 467.

(37.) A O B is a quadrant of a circle, O its centre; draw any chord B E D, and O E perpendicular to it, upon which take

O P equal to half the cosine of the arc B D, and determir the quadrature of the locus of P.

(38.) C P and C D are conjugate diameters of an ellipse, which the semiaxes are a and b; P F is perpendicular to D C: is required to find the area of the curve, which is the locus of

(39.) A right line drawn from a given point cuts a give circle, and the intercepted chord is the minor principal ax of an ellipse, whose area is equal to that of the given circl Find the quadrature of the curve described by the verte of the ellipse.

(40.) Supposing the arc of a semicircle to be stretched ot into a straight line, and an indefinite number of perpendicula erected on it, each equal to the versed line of the corresponc ing arc; what would be the length of the curve traced out b the tops of the perpendiculars?

(41.) The polar equation to the parabola is $r = \dfrac{a}{\cos^2 \dfrac{\theta}{2}}$; sho

that the area $= a^2 \left(\tan \dfrac{\theta}{2} + \dfrac{1}{3} \tan^3 \dfrac{\theta}{2} \right)$.

(42.) The equation to the lemniscate being $(x^2 + y^2)^2 = a$ $- y^2$, find its area between the limits of $x = 0$ and $x = 1$.

(43.) Let the base A B of a right-angled plane triangle b given, and in the variable hypothenuse A C, let there be con tinually taken C P equal to the perpendicular C B. Find th equation and quadrature of the curve, which is the locus o the point P.

(44.) A C B is a given quadrant of a circle; A the centre and D any point in the curve. Draw O D perpendicular to A B and take D P = B O; then the area of the curve, which i the locus of P, will be = the circular segment C D B C.

(45.) The perpendicular B C of a right-angled triangle A B (is given, and in the variable hypothenuse A C, let A P be taken, so as always to be equal to B C; required the distance B P of nearest approach, and the quadrature of the curve which is the locus of P.

(46.) Find the content of the solid generated by the revo- lution of the curve, whose equation is $(a^2 + x^2) y^2 - x^3$ $(a^2 - x^2) = 0$; about the axis of x.

G. Woodfall and Son, Printers, Angel Court, Skinner Street, London.

Lightning Source UK Ltd.
Milton Keynes UK
UKHW020659130922
408795UK00005B/306